YOUR AMAZING BODY

YOUR AMAZING BODY

HOW IT REVEALS GOD'S CREATIVE GENIUS

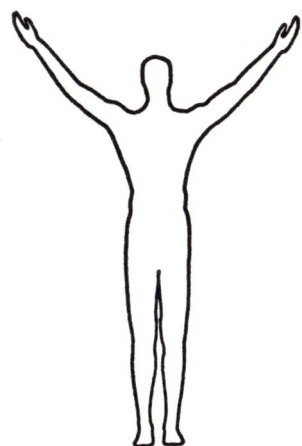

By Harold Shryock, M.D.

Southern Publishing Association, Nashville, Tennessee

Copyright © 1971 by
Southern Publishing Association
Library of Congress Catalog Card No. 76-155493
SBN 8127-0042-2

This book was
Edited by Richard W. Coffen
Designed by Dean Tucker
Text set in 11/13 Times Roman
Printed on Clear Spring Book Offset Hibulk
Cover stock: Springhill Gloss Bristol CIS

Printed in U.S.A.

CONTENTS

Introduction	9
The Wonder of Being Born	13
Your "Bring-and-Take System"	19
Your Bones Are Alive!	26
Your Wonderful Muscles	33
Skin Is for More Than Beauty	41
How Food Becomes Life	49
Your Incredible Liver	56
Keeping What You Need	63
Your Body's Subcontractors	70
How Life Begins	77
Temperature Control	83
Your Body's Defense System	90
Repair Work	96
Speech: A Marvel of Coordination	103
How Your Brain Knows	110
Your Brain Keeps Records	117
Movies of the Mind	124
Your Brain Is Boss	130
Healthy Body, Active Mind	138
How Faith Can Keep You Well	146
The Significance of Being a Person	153

INTRODUCTION

As a young child you took your physical body for granted. From your earliest recollections, you possessed a well-formed body with head, arms, legs, and the rest, and soon you spoke of "my fingers," "my foot," "my eyes," and "my mouth."

As you matured, your curiosity about your "insides" expanded. You observed that when the skin breaks, blood oozes; hence, you discovered blood inside your body. A continual craving for food, which depended a great deal on your activity, made you realize that food provides energy as well as the raw materials for growth. Thus, you became acquainted with the digestive organs.

Then you began to hear and read about other structures and organs: the heart, the lungs, the liver, the bones, the muscles, and the various glands. In school your teacher talked about your brain and helped you understand that it is the most marvelous of all your organs, for it houses the nerve cells which control the various parts of your body. Consciousness occurs in your brain. Your brain makes it possible for you to remember, imagine, and plan. Your brain makes your decisions. Your personality depends on what takes place in your brain. In your brain you relate to both man and God.

As your general knowledge of the human body has increased, you have become aware that each organ has its own function. Possibly you conceive of the body as just a collection of parts—like a mechanical device, with the heart serving as a pump, the liver playing the role of a chemical laboratory, and the kidneys taking care of waste disposal. Your reading of organ transplant exploits has augmented this idea of the body as a mere assemblage of organs.

But if your body is a machine, then it must have had a designer, since machines must be "invented" or "engineered" before construction. By examining a machine, a skilled mechanic can recognize what principles of engineering the designer used, and in some cases he can even guess the identity of the designer—simply because the item resembles other products by the same inventor.

Of course the human body is much more than a machine —with its complicated and incomparable intricacies. Yet upon your examining it carefully, two convictions emerge: First, it, too, had a Designer; second, this Designer possessed an intellect far exceeding the human brain. For years scientists have attempted to fathom the principles and natural laws that underlie the functioning of the human body. They have learned a great deal, but every new discovery seems only to introduce new and unsolved mysteries.

As we said earlier, when a child, you took your body for granted. You received it, complete in all details, with most of the parts already functioning. But who gave it to you?

The answer that first comes to mind is, Your parents gave it to you. In a biological sense this is true, for God endowed parents with the capacity to transmit life from themselves to the members of the next generation. But they cannot deliberately choose the kind of person their child will be or predetermine its sex. Parents serve as the agents through whom conception, early development, and birth take place as ordained by the Creator Himself.

Introduction

The tiny germ cells, one contributed by each parent, serve as miniature time capsules packed with the entire composite of inherited determiners. In terms of cell chemistry, giant molecules of deoxyribonucleic acid (DNA) form the actual transmitters of the hereditary code.

At the time of conception, when the germ cell from the father unites with that from the mother, a DNA molecule is established with an arrangement of atoms unique for that particular individual. This molecule functions as a blueprint to control the bodily structure and function of the unborn child.

As the embryo develops and the cells constituting the body of the new child increase, each new cell receives a "copy" of the DNA molecule which determines the hereditary pattern for this person. Thus every cell operates in harmony with the body's total scheme.

At first, following conception, all the cells composing the person-to-be are identical. Then, as development progresses, differences begin making their appearance. Certain cells take on the characteristics of young muscle cells; others, of nerve cells; others differentiate to bone cells; others, to the functioning cells of glands; and so on until all the body's fundamental cell types occur.

This remarkable process by which certain of the young cells "decide" to spend their lifetime producing saliva or some particular hormone or as blood cells with the capacity to ferry oxygen from the lungs to the tissues takes place very orderly, with no competitive gold-rush scramble on the part of the young cells for the more cherished opportunity to differentiate into brain cells. Just the right number of cells specialize as liver cells. Just the right number produce the strong fibers of connective tissue which hold the organs of the body in place.

What controls this remarkable process of cell differentiation? Biologists claim that the particular arrangement of atoms within the molecules of DNA controls differentiation. But how do the DNA molecules accomplish this? How do the DNA molecules form a unique molecular pattern for each particular person? Only those biologists who believe in God as Creator can answer these questions, and they base their answers on faith rather than on a comprehension of how God does it. These scientists share King David's sentiment of adoration: "Thy hands have made me and fashioned me." Psalm 119:73.

This concept of God's active power in the development and operation of the human body constitutes the prelude to an understanding of His personal intervention in the affairs of His created beings. The same God who brought you into existence concerns Himself with your physical and spiritual welfare throughout your lifetime. Again the psalmist clearly describes this relationship: "I praise thee for the awful wonder of my birth; thy work is wonderful. For thou didst form my being, didst weave me in my mother's womb. Thou knowest all about my soul, my body was no mystery to thee, as I was being moulded secretly and put together in the world below; all the days of my life were foreseen by thee, set down within thy book; ere ever they took shape, they were assigned me, ere ever one of them was mine." Psalm 139:13-16, Moffatt.*

---

*From The Bible: A New Translation by James Moffatt. Copyright 1954 by James Moffatt. Reprinted by permission of Harper and Row, Publishers, Incorporated.

THE WONDER OF BEING BORN

A ten-year-old boy, embarrassed over the subject of birth, says, "I don't remember anything about it."

An eighteen-year-old girl responds, "I think babies are wonderful, but I am not sure about motherhood. I understand that having a baby hurts."

A historian describes childbirth as "the greatest event in a person's life. Without it," he explains, "no other events could follow."

A philosopher says, "Birth introduces great uncertainties. For some it spells the beginning of a life of worthwhile accomplishments. But, alas, for too many it precedes misery, hardship, and disappointment."

A nurse manifests a typically human attitude. "Having a baby," she says, "causes some discomfort, to be sure. But modern ways of caring for a mother in labor remove a great deal of the suffering that took place a generation ago. And placing the new babe in its mother's arms awakens mother love so wonderfully that she soon forgets the pain."

A physician comments, "I prefer obstetrical cases. Such patients are not sick—just in need of professional care during one of life's most rewarding experiences. A mother and her husband are typically happy and appreciative individuals. A young mother in the prime of life responds well both physiologically and psychologically to professional care."

A minister recognizes in an infant a person-in-the-making—an individual whose intellect will develop, and with it the ability to choose and determine his way of life. The child will grow into a free moral agent responsible to God for the welfare of his own soul, and erelong he will wield an influence of his own.

A biologist views the birth of a baby as nature's crowning act. It represents the obvious transition between one generation and the next.

Yet birth embraces more than all this. Biologically speaking, the parents' bodies have been preparing for this event for many years. Even before birth the parents' sex organs contained a group of mysterious cells with the sole function of carrying the "spark of life" from one generation to the next.

The sex organs lie dormant through childhood, but with the advent of physical maturity, they make available these special cells charged with life-perpetuating characteristics.

Parenthood really begins nine months before the birth of the baby. Conception, when the highly specialized cells from husband and wife blend their life-giving constituents, initiates the development of the new individual. At conception the child's sex is determined. The equal blending of both parents' genes casts the die for the future child's personality.

This random selection of genes, with a different combination for each child born into the family, predetermines hair color, its nature (whether curly or straight), the eye color, skin tone, and the general pattern of body build (whether tall or short, stout or slender). These inherited traits produce a family resemblance, although no child exactly resembles either parent because the genes responsible for such inherited traits come from both parents.

The Wonder of Being Born

A child may inherit certain mental characteristics which bring advantage or disadvantage. It may inherit an aptitude for music but not the skill necessary for success as a musician. This the person must develop by individual effort.

After conception the number of cells that compose the body of the unborn child multiplies tremendously. Also, these cells organize into the various tissues and organs. In the first nine months the number of cells increases from the one combined cell contributed by the father and the mother to a total of 200 billion cells! A seven-pound baby at birth weighs two billion times what the original united cell at conception weighed.

The first three months after conception form the most critical period of physical development. During those three months the fetus must make many very delicate and complicated biologic adjustments as it outgrows the usefulness of certain temporary provisions for its needs and develops organs with long-range capacities.

Inasmuch as the early period of development is so important, the Creator saw fit to arrange an ideal environment for the first nine months of life—inside the mother's uterus, where amniotic fluid completely surrounds the unborn child. The fetal membranes enclose this fluid, and thus it serves effectively as a cushion to protect the growing body against mechanical injury. Also, the child can move about with relative ease while it exercises its arms and legs—even though the fluid-filled space crowds into tight quarters.

The heart of an unborn child develops quickly and actually begins to pulsate three weeks after conception. Hence, blood circulates early in the developing tissues.

Not so with the digestive organs, lungs, and kidneys. These develop more slowly and are not ready to function in time to care for the early, urgent requirements for nourishment, oxygen, and elimination of body wastes. Therefore, some of the mother's organs serve the baby's needs as well as her own. Food and tissue-building materials circulating in the mother's blood transfer to the blood of the child. Similarly, oxygen absorbed in the mother's lungs and carried by her blood seeps over into the oxygen-carrying cells of the child's blood. Likewise, the wastes that accumulate in its tissues transfer in the opposite direction, and the mother's organs of elimination expel them. Thus the child's tissues develop normally even before it can, on its own, capably care for certain of its biological needs.

At the beginning of the chapter we considered what birth means to persons in various walks of life. Now let us notice what it means to the child itself.

Childbirth greatly alters the processes within an infant's body. The most striking of these relates to the expansion of the lungs so that henceforth the child draws oxygen directly from the air rather than by an exchange of oxygen between the mother's blood and its own.

Should anything delay the child's ability to derive oxygen from its own lungs longer than three of four minutes after deprivation of oxygen from the mother's circulation, the brain cells would suffer permanent damage. In view of this, let us notice how the Creator has provided, even with a little time to spare, for a safe transition from total dependence on the mother to complete independence.

Prior to birth, the child's umbilical cord has carried blood between its body and the placenta, a temporary organ fastened to the lining of the mother's uterus in such a way that it permits the two bloodstreams (mother's and child's) to come into close contact, separated only by a paper-thin membrane. The twenty-two-inch umbilical cord is long enough to permit the unborn child to move freely within the

The Wonder of Being Born

fetal membranes and to pass through the birth canal while still deriving oxygen from its mother's blood. Beginning with the first breath, the newborn derives part of its oxygen from the mother and part from its own lungs for a few minutes.

As the lungs take over and blood begins to course freely through them, the amount of blood flowing through the umbilical cord gradually slackens. At this time, after the child has cried and thus taken a few deep breaths on its own, the doctor severs the umbilical cord. From now on the baby is truly an individual in its own right!

However, the infant's first use of its own lungs involves more than merely shifting from one source of oxygen to another. Before birth the route of blood circulation is adapted to the needs of an unborn child rather than to those of the postpartum infant. For example, before birth most of the blood bypasses the still-collapsed lungs via an opening directly through the partition that separates the right side of the heart from the left.

At the time of a baby's first cry, as the lungs expand, the pressures of blood within the various chambers of the heart change, thus permitting the soft tissues on either side of the aperture to touch and close the opening. Thereafter the pulmonary artery, which serves the lungs, receives as great a volume of blood as does the aorta, which supplies all other parts of the body. Here we see another wonder of childbirth.

Finally, let us notice what childbirth means to the mother. The capacity for motherhood requires many marvelous adaptations of the human body. The unique shape of the female pelvis, which now forms the firm support for the birth canal through which the child must pass, exemplifies these modifications.

However, the soft tissues within a woman's pelvis, at times other than childbirth, perform differing functions. Except at the culmination of pregnancy, the passage between the uterus and the outside is of such small proportions that it seems impossible it could accommodate an object even half the size of a baby's head.

But marvel of marvels, these soft tissues when relaxed and stretched permit the passage of a baby weighing between five and ten pounds; and if this were not marvel enough, following birth they return very nearly to their original size and position.

A different miracle occurs in the personality when a woman enters motherhood. Typically more sensitive, more responsive, and more tender in attitude than a man, the woman suddenly matures with great powers of stamina and endurance, coupled still with the qualities of mercy and patience.

The reality of motherhood makes a woman predominantly unselfish as she now gives first thought and energy without limit to the welfare of her little one. Personal sacrifice means little so long as it contributes to her baby's interests.

To this extent, childbirth brings a woman closer to God, and so all members of the family benefit because the mother's attributes in some measure reflect the divine character.

YOUR "BRING-AND-TAKE SYSTEM"

The human body is actually a community of living cells, each with its appointed function, and each with its own limitations. Of all these cells you might think a brain cell most fortunate. Located at the body's headquarters, it relays nervous impulses and also plays an important part in the control of other portions of the body.

Should this cell complain, it would have good reason to do so. It never goes anywhere, as the blood cells do. It can't contract and relax like muscle cells. It doesn't produce a secretion like the cells that compose a gland. On active duty about sixteen hours a day, it also goes "on call" during the hours of sleep. Being a specialist, this nerve cell merely passes on nervous impulses whenever the need arises. Yet even though highly specialized, it poses the same biological needs as the other normal living cells of the body. It requires food and oxygen, it must have an adequate supply of water, it needs certain vitamins and minerals, and it must have provision for disposing of the chemical substances left over from its vital functions.

It cannot go to the lungs to obtain the oxygen it needs. It cannot go to the digestive organs to get its food as a housewife goes to the market to buy provisions. Anchored in one location, it spends its life right there. For its own needs it must rely completely on the body's "bring-and-take system"—the cardiovascular system, as the scientists call it. Fundamentally the bring-and-take system consists of a constantly moving conveyor which serves every cell in the body except the constantly moving blood cells.

When we say "conveyor," we think of a moving belt or an escalator or a system of tubes such as carry sales slips and money in a department store. But obviously such a mechanical system would not work in the human body. The body requires a system that combines all the advantages of the mechanical devices just mentioned but with more versatility; in other words, a device that moves around almost every cell in the body, that carries all the ingredients a cell might need, that moves so quickly that it can supply emergency demands, that adapts to changing situations in various parts of the body, and that can carry away unneeded materials without having to make a second trip.

Who but the Creator could devise a system that meets such a formidable list of specifications? Yet it operates marvelously in every normal human body.

In the first place, the body's bring-and-take system is hydraulic—it consists of moving fluid, not of belts or cables. The fluid journeys throughout the body in a network of tubes: arteries, capillaries, and veins. The arteries carry the fluid (the blood) away from the heart, branching repeatedly to reach all parts of the body. The branches become smaller and smaller until they finally lead to capillaries—the smallest tubes of the bring-and-take system. The terminal ends of the capillaries unite with one another to form small veins, which carry the fluid back toward its starting point. Small veins flow together to make larger veins, and finally the largest veins open directly into the heart, which pumps the fluid throughout the system.

The capillaries come into intimate relation with the

Your "Bring-and-Take System"

body's various cells, such as the nerve cell in the brain mentioned at the beginning, and bring the cells what they need and carry away what they do not need.

Billions of capillaries, averaging about twenty-five thousandths of an inch in length, serve all the cells in each human body. Someone has estimated that if all the capillaries in one person's body were pieced together to make one continuous tube, the tube would reach more than twice around the earth.

In an average-sized person the bring-and-take system contains about six quarts of blood consisting of (1) plasma and (2) blood cells. Plasma is a straw-colored fluid composed mostly of water. Dissolved in the water are glucose (blood sugar), fat, various kinds of protein, minerals, hormones, enzymes, vitamins, antibodies, the waste products left over from the cells' activities, and small amounts of oxygen and carbon dioxide.

All the cells of the body have a constant need for oxygen. The uniting of oxygen with food materials within the cell provides the energy for the cell's vital functions. This production of energy forms carbon dioxide, which is, as far as the cell is concerned, waste material; and quick removal of carbon dioxide makes room for more oxygen within the cell.

The small amounts of oxygen and carbon dioxide dissolved in the blood plasma are insignficant compared with the large quantities involved in maintaining the chemical processes within the cells. So the Creator has provided a special carrying device within the blood to transport huge quantities of oxygen and carbon dioxide—the humble red blood cell containing giant hemoglobin molecules.

Hemoglobin, a complex protein containing iron, accounts for the red color of the red blood cells, and these, in turn, for the red color of blood. A hemoglobin molecule has the magic ability to latch onto a molecule of oxygen. This union between hemoglobin and oxygen is frail, yet just strong enough, under prevailing conditions, to carry the oxygen from the lungs to the cells in outlying parts of the body.

Red blood cells also carry 95 percent of the carbon dioxide back to the lungs. Chemically, the means by which red blood cells carry carbon dioxide differs from the means by which they carry oxygen. Suffice it to say, they get it back to the lungs, which eliminate it from the body through the expired air.

We have difficulty realizing the small size and vast numbers of these red blood cells, shaped like doughnuts with their thin centers not quite punched out. It would take 3,500 of these cells placed side by side to make a line one inch long. The blood fairly teems with these tiny "packages" for carrying oxygen and carbon dioxide. A drop of blood contains about 300 million of them. The average red blood cell lives a little more than one hundred days. But nearly 30 trillion of them float in the blood of an average-sized adult. This means that in order to replace those that wear out, the body's bone marrow, which produces the red blood cells, must turn out more than three million new cells every second.

An adequate supply of blood to all tissues, regardless of the position of the body, is ensured by the continuous activity of the heart, which functions as a pump. The blood does not flow by gravity but is forced under pressure to all parts of the system, whether in the head or in the feet.

A principle of hydraulics states that the resistance to a flow of fluid maintains pressure. Here the small caliber of the capillaries provides the resistance. The capillaries are

Your "Bring-and-Take System"

so narrow that in some places the blood cells have to squeeze through single file. The amount of work which the heart normally performs in maintaining pressure within the arteries equals that required to lift a ten-pound weight three feet off the floor every thirty seconds.

It averages less than half a minute for a drop of blood to leave the heart, pass through the arteries, capillaries, and veins and return to its starting place. A total volume of more than 4,000 gallons of blood passes through an adult heart daily.

The walls of the arteries stretch easily. Blood pressure would fall to zero between one heartbeat and the next in rigid arteries. The arterial walls contain a large amount of muscle and elastic tissue. Therefore each time the heart pumps a new volume of blood, the arteries expand. Then, between heartbeats, the arteries come back to their previous size, maintaining as they do so a certain degree of pressure within the system.

Now we can understand why the doctor writes down two figures when he takes a patient's blood pressure. The larger figure represents the pressure produced at the peak of the heart's contraction. The lower figure measures the pressure remaining in the system between heartbeats. The pressures are expressed, the same as in measuring barometric pressure, in terms of the height of a column of mercury that such a pressure would support. The doctor usually writes the two figures together, as 120/80, which, incidentally, is a normal blood-pressure reading.

The bring-and-take system can remarkably adapt to meet the changing demands of time and place throughout the body.

Blood pressure is greatest, of course, in the aorta—the giant artery leading away from that part of the heart which produces the highest pressure. As the aorta branches and as these branches continue to divide and discharge their blood into smaller and smaller vessels, the blood pressure reduces progressively, as does the speed of the flow of blood.

Finally, as blood reaches the capillaries, it moves so very slowly that it requires about two seconds to pass through a capillary as long as the thickness of twelve sheets of paper. At this slow speed the red blood cells easily give up their cargo of oxygen molecules and take on the waiting molecules of carbon dioxide.

Blood pressure in the veins is very low—so low that in some veins it registers as a negative pressure. Simple valves in the walls of the veins help the blood back to the heart. These keep the blood from flowing backward. Therefore, any movement of blood within the veins must flow in the direction of the heart. As a person moves the large muscles of his body, the expanding muscles squeeze the veins and thus force the blood on its way.

Blood pressure throughout the body and the rate of blood flow at any given point in the system vary from time to time in harmony with the body's needs. Here we have another evidence of the marvelous design of the human body and its various systems.

When a person exercises, his muscles must have a much greater volume of blood to provide the additional oxygen and blood sugar that the increased activity requires. The nervous system arranges the details by increasing the rate and force of contraction of the heart, which thus pumps more blood in a given period of time. The heart has a tremendous capacity for increased work and can do ten times as much work in an emergency as it does ordinarily when the body rests.

But merely increasing the work of the heart does not of

Your "Bring-and-Take System"

itself ensure a maximum supply of blood to hardworking muscles. So the nervous system arranges yet another adjustment, this time by restricting the flow of blood to parts of the body not involved in the muscle activity of the moment. By tightening the delicate muscles in the walls of the arteries, the arteries constrict so that they offer more resistance than usual to the flow of blood through their lumina.

In principle, the same thing occurs after a meal when the organs of digestion need a greater-than-usual supply of blood. Also, when a person actively engages in study, the bring-and-take system adjusts to carry a large volume of blood to the brain even at the expense of the amount of blood flowing to other parts of the body. This explains the cold feet of which some diligent students complain.

It would be hard to decide which of the body's systems or organs is most important or most remarkable in its design and control. Each has its own work to do, and each fits into the total scheme of things in a way that makes it indispensable. Each system contributes to the total welfare of the human being.

Reviewing this amazing bring-and-take system, which so ably illustrates the Creator's handiwork, one can but exclaim with the psalmist, "I will praise thee; for I am fearfully and wonderfully made: marvellous are thy works; and that my soul knoweth right well." Psalm 139:14.

YOUR BONES ARE ALIVE!

"Why is it that my wife's broken hip refuses to heal as it should?" an elderly gentleman asked the doctor.

"The break occurred in an unfortunate place," the doctor explained. "In this case a fragment broke away from the remainder of the bone and has no blood supply of its own to trigger healing."

"You mean the broken piece of bone is dead?"

"Well, you might say it that way," the doctor agreed.

"I hadn't realized that bones are alive in the first place," the man admitted. "I had supposed that bones resemble the steel framework of a building."

"They do," the doctor went on, "but they have many other qualities, too. If bones were not composed of living tissue, how do you suppose they grow during childhood—to say nothing of their usual ability to repair themselves after a fracture?"

"I just hadn't thought of it that way," the man acknowledged.

Bones consist, as do other tissues of the body, of living cells plus supporting material produced by those cells. In the case of bone, the supporting material occupies more space than the cells, but the cells, present throughout life, will produce more of the supporting material should the need arise. Other bone cells (osteoclast cells) have the ability to destroy the supporting material as necessary in the process of changing a bone's architecture at various times of life.

The supporting material located between bone cells consists of collagen in the form of many closely packed fibers. Collagen, the same material found in tendons, has great tensile strength. Minute crystals of the salts of calcium and phosphorus fill in among the collagenous fibers, making the bone firm and rigid. The blood brings these molecules of mineral salts to the bone and deposits them there.

As bone-forming cells produce collagen, they actually become imprisoned in the substance of their own making. Each bone cell thus resides within a tiny space surrounded on all sides by supporting material. But the imprisonment is not absolute, for tissue fluids, which bring food material and oxygen and carry away waste products, can adequately reach each imprisoned cell. Whereas the cell cannot move from place to place, a system of minute canals (called canaliculi), which pass in various directions through the bone substance and carry the tissue fluid that has seeped through the wall of the nearest blood vessel, provides all its vital needs.

Blood vessels abound in and around every bone of the body, and they do not run past the front doorstep of every bone cell, but they pass through special channels into the interior of all bones. Each long bone has a central cavity which runs from one end to the other. A few blood vessels pass through the dense wall of such a bone, enter its central cavity, and form many branches. From these, branches of capillary size penetrate the bone substance and supply the tissue fluid, which flows slowly through the canaliculi that serve all the imprisoned bone cells.

The flat bones of the body do not have wide-open central cavities but rather an internal honeycomblike structure in which the penetrating blood vessels branch and rebranch, thus supplying the needs of the imprisoned bone cells located within the bony partitions of the honeycomb.

A membrane known as the periosteum covers all bones, whether long or flat. Blood vessels present in this membrane provide another source of tissue fluid to sustain the entrapped bone cells. The branches of these vessels in the periosteum penetrate the bone substance, beginning externally and meeting, at about the halfway point, the blood vessel branches that pass outward from the bone's interior.

One of the most convincing evidences that live tissue makes up bones is their ability to increase in size during the period of growth. When a contractor enlarges a building such as a hotel, he must add more rooms to those already present and the original rooms remain the same size. But when the bone grows, all of its proportions increase. It becomes heavier, longer, larger around, and thicker.

When a building is being enlarged, the normal activities that take place in the building have to be curtailed until the completion of all alterations. Not so with a bone. It retains its ability to support the body all the while it grows.

Picture in your mind the size of the bone in the upper arm (the humerus) of a newborn baby, and contrast this with the same bone in the arm of a grown man. Between infancy and adulthood the bone has grown both in length and in circumference. The central cavity of the bone has so enlarged that the entire bone of the baby would fit into it.

The growth of a bone does not consist of a mere stretching of its tissue elements, for bone tissue is firm and stout and cannot stretch. During growth additional bone tissue develops. Plastering new bone on the outside of the old would not suffice, for the size of the central cavity must increase at the same time the bone's circumference enlarges. Neither would adding new bone to the surface of a baby's

Your Bones Are Alive!

bone make it sufficiently longer or adequate.

The growth of a bone such as the arm bone requires three separate, intricate processes, occurring simultaneously.

First the bone lengthens. Taking the arm bone as an example of the body's long bones, we find a shaft with an extremity at each end. The shaft roughly resembles the shape of a simple tube, while the extremities are shaped in ways that adapt them for being components of the shoulder joint and elbow joint respectively.

Throughout the years of childhood and youth the extremities of the arm bone do not as firmly adhere to the shaft as in adulthood. During the growth period, a thin plate of unique tissue which produces new bone interposes between each extremity and the shaft. This new bone is added to the shaft, making it longer. Called the epiphyseal cartilage, this plate of bone-producing tissue disappears in the late teens or early twenties. Thereafter the bone cannot increase in length.

The second process involved in bone growth consists of the addition, a layer at a time, of new bone around the outer surface of the one that already exists. We can compare this to the increase in circumference of a tree trunk as a new ring of wood is added each year, although in the case of bone, the layers grow faster than one a year. Also, the adding of new layers stops at adulthood.

The third process in relation to the growth of bone, not as spectacular as the other two, is fully as important in what it contributes to the development of a properly designed adult structure. This involves the destruction of previously formed bone so as to enlarge the central cavity of the shaft. As bone accumulates on the outside surface and ends of the growing bone, specialized cells in the interior dissolve away those portions no longer needed. This not only makes room for larger blood vessels as needed within the bone but helps carry out the Creator's policy of arranging for maximum strength with a minimum of weight.

We have mentioned that the flat bones, such as those which protect the brain, do not contain a central cavity but, instead, have a honeycomblike interior. This holds true, also, for many of the irregularly shaped bones and for the extremities of the long bones. Comparison with a honeycomb does not give an accurate impression, however, for the arrangement of the bone components more closely approximates girders in a steel trusswork than the partitions of a honeycomb. In fact, comparison with a steel trusswork is quite helpful, for the internal architecture of these areas of bone provides the greatest possible support along the lines of stress.

And now we come to another evidence that bone is truly a living tissue. The architectural pattern of the minute girders of bone that occurs internally constantly alters as a continuing adaptation to the demands of the activities in which the individual engages. As certain muscles increase in strength, they pull harder than before on the bones to which they adhere. With this as a stimulus, bone-forming cells build new bone to give internal reinforcement where necessary. Bone-destroying cells may remove fragments of bone that no longer function or that stand in the way of the new architectural pattern. This reorganization of bone structure continues even during adulthood.

Bones come into contact with each other at the joints. The movement of one bone in relation to another makes it possible for a person to change the shape and position of his body and its parts. The ends of the bones as they contact each other in a joint serve as bearings do in a mechanical device.

Your Bones Are Alive!

Of course, bone tissue is hard and quite unyielding, and bone in direct contact with bone in a joint would have a scouring, abrasive effect, and the contacting ends of the bones would gradually wear away. But here again we find evidence of the Creator's design, for a thin layer of durable and resilient cartilage coats those surfaces of the bones which contact each other at a joint (the articular surfaces). The cartilage surface of one bone contacting the cartilage surface of the next bone smooths joint actions and makes them precise.

Bone marrow fills the central cavities of the long bones, the spaces of the honeycomb arrangement in the heads of long bones, and the inner central parts of flat bones—tucked away in otherwise empty and useless spaces. Here in its out-of-the-way locations marrow receives a generous supply of blood, enabling it to carry on its important function of producing most of the blood cells. As blood flows through the marrow, the newly formed blood cells cast off from their parent cells and float away, spending the rest of their life-span as a very important part of the blood.

However, blood cells do not live forever. Red blood cells—the most numerous—live only between two and three months. Thus the bone marrow must continuously produce red cells and many of the other kinds of blood cells which help maintain the body's defenses.

Early life taxes bone marrow almost to the limit of its capacity, to produce the increasing number of blood cells which the growing body requires. Of course, as the size of the body increases, the amount of marrow tissue increases proportionately. Then at adulthood growth halts, the demand for new blood cells levels off, and some of the once-active marrow tissue now serves only a standby function.

This less active marrow is called "yellow marrow," for

it now contains a considerable number of fat cells and only scattered islands of blood-cell-producing elements. But even though usually quiescent, it remains an important tissue, for in case of sudden or unusual demands for new blood cells—as following the loss of blood by hemorrhage—yellow marrow may resume its function of producing blood cells and thus supplement the work of the continuously active "red marrow."

In discussions of the marvels of the Creator's handiwork as it relates to the human body, we frequently select the brain or the heart as the part of the body which we consider most remarkable and which functions in a way almost beyond comprehension. But God perfected His work in every detail, and the more attention we give to any part of the body, the more wonderful it seems. Even the lowly bones, as they have just been described, reflect a design and a degree of forethought that forces a new appreciation and a recognition of an All-wise God.

Each tissue of the body—even those tissues which compose the bones—has its own vital function. In a broader sense, design and purpose embrace every unit of God's creation. God has a plan for the life of each human being—a plan which reaches its most glorious fulfillment only as the individual chooses to relate himself to God and to his fellowmen in harmony with the code of living which He clearly presents in His message—the Holy Scriptures—to the human race.

YOUR WONDERFUL MUSCLES

If, as someone said, life is motion, muscles must be the essence of life, for they, as far as the animal kingdom is concerned, make motion possible. Muscles compose more than 40 percent of the human body.

Cardiac muscle in the wall of the heart produces the pressure to propel the blood. A different kind of muscle (smooth muscle) in the walls of the blood vessels provides control of the blood pressure and directs the greatest flow of blood to the part of the body which needs it at the moment.

Skeletal muscle in the diaphragm and between the ribs enables one to breathe. The same kind of muscle in the larynx, the diaphragm, and the wall of the abdomen makes it possible to talk, sing, or scream.

Muscles in the tongue and face provide for chewing food. Muscles in the tongue, pharynx, and esophagus cause swallowing after mastication. Smooth muscle in the walls of the stomach and intestine churn the food during digestion and move it on its way along the alimentary tract.

Smooth muscle in the framework of the spleen permits this organ to force out an extra pint of blood into the general circulation following a hemorrhage.

Delicate strands of smooth muscle in the eye alter the shape of the lens in order to properly focus light rays. The same kind of muscle in the eye regulates the amount of light that enters through the pupil. Tiny skeletal muscles, precisely placed and accurately controlled by coordinating centers in the brain, move the eyes to enable them to provide a view of whatever a person wants to see.

Miniature skeletal muscles in the face register moods and emotions by changing facial expression. Tiny smooth muscles placed in the skin at the roots of the hairs cause goose pimples and permit the hairs to "stand on end" in cold weather or during intense emotions.

Skeletal muscles in the neck and shoulders permit a person to move his head as he pleases. Muscles in the back, shoulder, arm, forearm, and hand enable a person to play a musical instrument. It takes a careful coordination of the skeletal muscles of various parts of the body to lift a person out of bed, to bring food to his mouth, to keep his body erect while he moves one foot after the other in walking. Driving a car, water-skiing, and operating a power lawn mower involve the same kind of coordination and skeletal muscle action.

Truly, muscles are necessary to life! At any given moment many of the body's skeletal muscles serve a standby function, just waiting for some need to arise which will call them into activity. Not so with the highly specialized cardiac muscles in the walls of the heart. Not only do the cells of this muscle possess the same ability to contract that characterizes the cells of all kinds of muscle, but they also have the added ability to contract without being told. They do not wait for stimulation from the nerves, but automatically initiate one heartbeat after another, about 100,000 times a day throughout life.

Even so, the nerve impulses do influence cardiac muscle. But for these, it would continue indefinitely to beat at a constant rate and a uniform strength of contraction. But the nerves bring impulses which modify the rate and strength of the heart's action, making it faster and more forceful when blood is in short supply throughout the body or allowing it to go back "on automatic" upon reduction of the body's needs. And here we see demonstrated the marvelous capacity of cardiac muscle to respond to sudden demands for increased activity.

When a person rests, his heart beats about sixty times a minute and pumps about four quarts of blood. The heart responds to strenuous exercise by increasing its rate and pumping as much as sixteen quarts of blood in a minute. During extreme exertion, as by a trained athlete, the heart may pump as much as thirty-five quarts of blood per minute for a short time.

Next let us notice how smooth muscle is continuously ready for action.

Smooth muscle does a very different kind of work from that performed by cardiac muscle. By comparison, smooth muscle operates slowly and deliberately. Its automatic contractions may occur as often as every five seconds or as infrequently as every few minutes, depending on the location and function of the particular sample under consideration.

Smooth muscle—found, for the most part, in the walls of the body's tubes (such as blood vessels, ureters, and intestine) or in the walls of the hollow organs (such as stomach, gallbladder, urinary bladder, and uterus)—can contract with about equal force regardless of the present size of the organ in which it is located. For example, the smooth muscle in the walls of the stomach can just as actively agitate a small amount of food within the stomach as when the stomach is filled to its capacity.

Nerve impulses exert even more influence upon smooth

muscle than upon cardiac muscle. In the walls of such organs as the stomach, the intestine, and the bladder, small collections of nerve cells locally influence the smooth muscle and control many of its actions. For example, when digested food leaves the stomach and moves into the intestine, its very presence stretches the intestinal walls and sets off local nerve reflexes which cause the muscle to churn the contents and propel them gently on their way.

Nerve impulses from the autonomic nervous system also influence smooth muscle. Some of these cause the muscle to contract, and others, to relax.

When a person is alarmed, his autonomic nervous system sends out signals which cause the smooth muscles in the digestive organs to stop functioning but at the same time stimulate the muscles in the walls of the blood vessels to contract and thus raise the blood pressure. Once the crisis ends, the muscles in the walls of the arteries relax and those in the digestive organs again take up their rhythmic contractions.

Perhaps our body's skeletal muscles furnish our best example of muscle being always ready for action. In most locations these muscles adhere to bones, and they principally function to move these bones, thus permitting changes in position and bringing about movement from place to place.

Skeletal muscle, under the definite control of the nervous system, has to be stimulated by a nervous impulse in order to contract. On this account it is sometimes called voluntary muscle.

The individual cells (often called "fibers") of skeletal muscle provide the power to move the bones of the body. Each of these cells has its own connection with a tiny nerve fiber and contracts only as the nerve fiber conveys an initiating impulse.

Your Wonderful Muscles

Even though the individual fibers of skeletal muscle are only about as large in diameter as a human hair, they reach lengths of ten inches in some of the body's larger muscles. In the smaller muscles the fibers are proportionately shorter. The fibers run lengthwise in the muscle and attach at each end, either to another muscle fiber or to the filaments of a tendon anchored to some bone of the skeleton. Thus when muscle fibers contract, the entire muscle shortens correspondingly, and the bone to which the tendon fastens moves.

Such joints as the elbow and knee typically include two sets of muscles—one set which flexes or bends the joint and the other which extends or straightens it. The two sets obviously have opposite actions, and if both sets contract with equal force, no motion results. To flex the elbow, the extensor muscles must relax at the same time the flexor muscles contract. Similarly, in order to straighten the elbow or the knee, the flexor muscles must relax as the extensor muscles contract. Control centers in the spinal cord effectively care for this reciprocal action. When a person knows that he wants to bend his arm at the elbow, he does not have to give attention to relaxing the muscles that would oppose this action. The nervous system automatically cares for it all.

In order for the parts of the body to move smoothly, opposing muscle groups must never relax so much that they permit their tendons to become slack. Should this occur, the movements would awkwardly jerk. During all waking hours, then, the various skeletal muscles of the body maintain some firmness even though they are not actively contracting. This firmness, when supposedly at rest, helps to keep the muscles constantly ready for action.

Suppose you desire to lift a book from the table. The amount of strength required will depend on the weight of the book, a heavy book requiring more lifting power than a light one. But the same muscles do the job whether the book is heavy or light.

The amount of power a muscle exerts when it contracts depends on more than one factor. The factor of most importance, however, is the number of the muscle's fibers that actually contract. When you lift a light book, relatively few of the fibers contract, and these are scattered here and there throughout the muscle. The contraction of these few fibers causes the entire muscle to shorten, but it does so gently. When you lift a heavy book, a proportionately greater number of the muscle's fibers go into action. In an all-out effort to lift a very heavy weight, all of a muscle's component fibers contract.

Suppose you wish to hold a book for a while as you read. This requires a sustained activity in the muscles concerned. Even muscle fibers, powerful as they are, become tired during prolonged contraction. In a case such as holding a book while reading, not all of the fibers in the muscle contract at any one time. So, the fibers function in relays, certain ones being active while others rest.

The mild firmness that keeps muscles taut and ready for action even when not in use occurs in about the same way. The nervous system apparently delivers a few nerve impulses to each of the skeletal muscles at all times—just enough to keep a small portion of their fibers in a state of contraction.

You probably know already that the energy required for the action of your muscles derives in the long run from the food you eat. We hear about the process of oxidation within the body in which blood sugar in the tissues unites with oxygen to produce energy in a way similar to the burning of fuel in an engine.

But when it comes to the movements of muscles, there

is not time to wait for the blood sugar to combine with oxygen and thus provide the necessary energy. The energy must be available right now without waiting to find some molecules of blood sugar and introduce these to some molecules of oxygen.

The human body has many reserves by which it keeps in store more than enough of each of the substances it may need. It can draw on these reserves on short notice at any time and so carry on its functions, even in an emergency.

So it is with the storage of energy which makes possible the contraction of muscles. The oxidation of blood sugar within the tissues brings about a chemical reaction which produces a highly energized substance called ATP (adenosine triphosphate), which the muscle cells store.

When a muscle cell receives a nerve signal to contract, energy suddenly releases from the ATP molecules. This causes a slight change in the two kinds of protein (actin and myosin) which are always present in muscle fibers and exist there in the form of tiny filaments. These protein filaments lie endwise within the muscle fiber with the ends overlapping slightly; but when the fiber relaxes, the actin and myosin filaments do not lie side by side. Once they become activated by the release of energy from ATP, their newly acquired electrostatic charges impel them to slide past each other—as you can illustrate by sliding the extended fingers of your right hand into the spaces between the extended fingers of your left hand. This action, occurring among all the filaments within a muscle fiber, shortens the total muscle fiber and thus tugs on the tendons which attach it to bone.

When a muscle fiber contracts, its supply of ATP, of course, diminishes. The supply is restored at leisure by a process which combines blood sugar with oxygen.

Thus we see that, throughout the body, muscles provide

the power for motion, whether it be needed to propel the blood, to digest one's food, to raise an eyebrow, or to drive a golf ball. But the provision by which these muscles are constantly in readiness for action as occasion requires seems even more marvelous than the presence of specialized muscles in strategic locations. Truly the wonders of the human body justify our continuing adoration of the Creator!

SKIN IS FOR MORE THAN BEAUTY

When the teacher asked Johnny the purpose of the skin, he replied, "The skin keeps the body from looking raw."

True enough, the skin does serve as a barrier between the atmosphere which surrounds the body and the tissue fluids within. It seals off the body fluids so that little evaporates. To this extent, Johnny answered correctly.

As the body's external barrier, the skin prevents trespassing in either direction. Not only does it minimize the escape of body fluids but it also prevents the entry into the body of many possible noxious substances, thus establishing the body's first line of defense against the entry of irritating dusts, chemicals, and germs.

The skin, however, is much more than a tight-fitting plastic bag which protectively surrounds all parts of the body. It helps control the body's temperature. It constitutes an important sense organ, making use of touch, a sensitivity for heat and cold, a provision for registering pressure, and a signal system producing pain to help keep a person aware of his surroundings. The skin contains numerous small glands which discharge their fluid or oil onto the body's surface. Finally, the skin repairs itself following injury.

In order to understand how the skin can accomplish what it does, we must know its construction, for it is built very precisely in a manner adapting it to play its various roles.

No single description applies to the skin of all parts of the body. Its features in one location differ from those in another, modifying it for protection against heavy friction on the palms of the hands and soles of the feet, for flexibility around the joints and parts of the body which move freely, for remarkable sensitivity around the face and other parts of the body where sense reception is important, and for the growth of hair in those areas where hair provides added protection or improves a person's appearance.

In all parts of the body, however, the skin has two principal layers—a surface layer of closely packed cells, making it virtually waterproof, and a deeper layer of strong fibers arranged in a feltwork, lending it both strength and flexibility. Scientists call the surface layer the epidermis and the deeper layer the dermis or corium.

The epidermis contains no blood vessels or nerves—just layer upon layer of closely packed cells. The deepest of these cells are very much alive and actively produce more cells just like themselves. As more cells develop, they crowd their way toward the surface, meanwhile moving farther away from their source of nourishment and undergoing chemical changes, thus turning horny and scalelike. When they reach the surface, they are no longer alive and eventually rub off by friction with the towel after bathing or by coming in contact with some unyielding surface.

The thickness of the epidermis increases in those parts of the body exposed to wear and tear. A child walking barefoot acquires a very thick epidermis on the soles of his feet. An older person whose shoe pinches develops thick areas (calluses) at the points of contact with his shoe. A violinist forms a thick epidermis on those parts of his fingers which

Skin Is for More Than Beauty

press against the strings. In all persons the skin of the eyelids has such a thin layer of epidermis that it is very sensitive to touch.

The translucent epidermis would be more nearly transparent were it not that its flat, scalelike surface cells are more opaque than the deeper cells. Even so, the rosy color of the blood in the small vessels of the underlying dermis shines through in certain parts of the skin, imparting its characteristic pink hue.

The deeper cells in the epidermis may contain tiny granules of pigment—much more so in a dark-complexioned person than in a blond. The important pigments here are melanin and carotene—the more melanin, the darker the complexion. The carotene imparts a yellowish hue. Dark skin contains so much melanin that the epidermis loses its translucence.

Now we turn our attention to the deeper layer, the dermis. As already mentioned, the foundation material of this layer consists of a feltwork of strong connective-tissue fibers. In any given area of skin the fibers tend to run all in the same direction, thus providing strength in the line of the greatest mechanical stress for the area. This accounts for the direction of the tiny folds that appear in the skin, more prominent in some areas than in others.

The dermis houses the structures that provide the skin's "utilities." Here we have nerve fibers and blood vessels of various sizes. The dermis and the loose connective-tissue layer just beneath it hold the skin glands, the receptor organs for the various sensations, and the functioning portions of the hair follicles which produce and maintain the hair shafts. Many of the capillaries (the smallest blood vessels) lie very close to the boundary between the dermis and epidermis. This not only permits the color of the blood to shine through the epidermis, but allows some of the fluid portion of the blood to seep into the epidermis and thus nourish its active, deeper cells.

The nervous system rigidly controls the flow of blood through the vessels contained in the dermis. When the body is in cold surroundings, little blood flows through the skin— just enough to provide nourishment for the skin cells. But when extreme heat surrounds the body, the amount of blood flowing through the skin increases as much as one hundred times. This is the body's principal means of eliminating heat to keep internal temperature relatively constant.

As this large volume of blood flows through the now dilated vessels of the skin, the skin warms, and its reddened appearance offers evidence that much blood flows just beneath the epidermis. Under these conditions, the sweat glands produce sweat, which flows through small openings in the epidermis onto its surface, where it evaporates. This evaporation cools by dissipating the heat which the blood brings from the body's deeper tissues. Under conditions of extreme heat, the sweat glands can seep several quarts a day.

The fluid which the sweat glands exude consists mostly of water, but it also contains a small amount of ordinary salt (sodium chloride). Excessive sweating, then, heavily taxes the body's supplies of water and salt. For this reason a person exposed to high temperatures should drink large amounts of water and, in extreme cases, should take salt tablets to prevent the heat cramps which might otherwise ensue.

The boundary between the dermis and the epidermis does not follow a straight line or a smooth surface. The underlying dermis has many hills and ridges which fit into corresponding pockets and grooves on the under surface of the epidermis. This arrangement provides a firmer attachment

Skin Is for More Than Beauty

for the epidermis than if it were merely placed over the dermis like one sheet of paper on another. Even so, when some object rubs part of the skin too vigorously, as when a person with soft hands keeps gripping the handle of a tool, the epidermis may separate and permit the formation of a water blister.

The prominent ridges and grooves in the skin of the palms and soles account for fingerprints and footprints. Their particular pattern—unique for each individual—serves as a means of personal identification and provides a nonskid quality, increasing the firmness of a person's grip with his fingers and the surety of his step when walking barefoot.

A person leaves finger marks on a pane of glass, or on a polished surface, because the sweat glands pour out their secretion through tiny openings located along the ridges of the fingerprint pattern. The sweat glands abound in the skin of the palms, soles, face, and neck.

Certain parts of the body, such as the armpit and groin, have larger than usual sweat glands associated with the roots of the hairs in these areas. Slight discoloration and a distinctive odor mark the secretion of the sweat glands in these locations.

Along the margin of each eyelid a single row of highly specialized sweat glands secretes a considerable amount of oil. The film of oil thus produced serves two important purposes. First, it seals the eyelids during sleep to prevent the evaporation of the fluid produced by the tear glands which keeps the eye structures moist. Second, it prevents the overflow of the fluid during waking hours, which usually is produced continuously in moderate amounts. After flowing between the eye and the inner surfaces of the eyelids, it passes through the tear ducts at the inner corner of each eye and into the corresponding nasal cavity. The film of oil at the margin of each eyelid prevents the fluid from overflowing onto the cheek rather than passing through the tear duct. Of course, when a person cries, the volume of fluid produced by the tear glands so increases that it overflows anyway.

Associated with the roots of the hairs, sebaceous glands produce an oily substance that keeps the skin pliable and prevents the hairs from drying out.

In connection with each hair root, and thus with its associated sebaceous gland, a very tiny muscle lies buried in the dermis of the skin. This so-called erector muscle responds to extremes of cold and even to fright. When the erector muscles contract, the hair roots move into a more perpendicular relationship to the skin surface, thus causing the hairs to "stand up." Because the muscles lie right in the skin, their contraction causes "goose pimples." Incidentally, the contraction of each of these tiny erector muscles makes pressure against the corresponding sebaceous gland, causing it to discharge its oily fluid along the root of the hair.

In the dermis, near where it contacts the epidermis, nerve fibers connect the tiny sense organs to the brain, where the sensations of touch, of heat and cold, of pressure, and of pain register. A different kind of sense organ exists for each of these sensations. Each specializes by producing only its own kind of sensation, and careful manipulation with a needle point can stimulate these organs one at a time. Even though the needle is of neutral temperature, when it contacts a sense organ designed to receive sensations of heat, the needle point "feels" hot. Similarly, it "feels" cold when it contacts a sense organ for cold.

Perhaps a person can best appreciate how the skin provides important sensations by noticing what happens when

Skin Is for More Than Beauty

he prowls about the bedroom in the dark. He cannot see because of the darkness. He hears nothing because all is silent. But the sense organs in the skin still function at full capacity and provide him with the information he needs at just such a time.

He gropes his way from one piece of furniture to another by making contact, usually with his fingertips, with familiar objects. If he tends to fall, pressure produced by the side of the bed or by the arm of a chair stimulates his sense organs for pressure, thus informing him about the object he needs to avoid. If, carelessly, he stubs his toe, sensations of pain quickly reprimand him for moving too quickly. If the room is cool, the receptor organs for cold notify him of this, and he arranges for another blanket on the bed. After his prowling, he returns to the warmth of his bed, and the organs for heat, as they are moderately stimulated, give him a sense of comfort. But if, in the meantime, he has turned on the heating pad, these same organs for heat will warn him if overheating threatens.

The skin's delicate tints, its waxy appearance because of the translucent epidermis, its capacity for fitting neatly even when the body moves, and its ready adaptation to changes in position and in facial expression—all have eye appeal and contribute to our awareness of beauty.

The Creator endowed mankind with an appreciation of beauty. Furthermore, He provided for our enjoyment thousands of items of natural beauty. Therefore, we should cultivate an appreciation of the beautiful and allow things of beauty to lift our souls and attract our thoughts to lofty themes.

But beauty for beauty's sake is not a worthy life goal. Physical beauty, so often only skin deep, may fade with the passing of time. Beauty of character is more than skin deep, however, and will endure year by year, even into eternity. As a person cultivates those personal traits which make him generous, forgiving, patient, and unselfish, his friends acclaim his "beauty" whether or not the skin which covers the body tickles the eye.

HOW FOOD BECOMES LIFE

We recognize that a man must eat to live, but most people take for granted how eating makes living possible. As scientists have discovered what happens to the molecules of food entering the workings of the human body, they have revealed another marvel of creation and new evidence that divine forethought made it possible for us to "live, and move, and have our being." (Acts 17:28.)

Of course we human beings can go right on eating and living whether or not we know the details of how food satisfies the body's demands for energy fuel and at the same time provides the materials for growth and repair of the tissues. And we continue to thrive on the food we eat, even though uninformed on how the body utilizes it, because we have a built-in craving which forces us to want food regularly every day. This appetite for food keeps us from neglecting to take what our bodies need to maintain life.

Fortunately, we get tired of the same kind of food. Thus we vary our diet from meal to meal and from day to day, which helps save us from the danger of a deficient diet, for what one food may lack in needed nutrients, some other food contains.

A person's appetite for certain foods depends largely on his tastes. One's choice of food at a certain meal is governed more by this factor of taste than by his knowledge that the food contains a certain type of protein or that it provides a valuable vitamin. However, the use the body makes of a certain food depends not on its taste but on the food constituents it contains.

A certain piece of cloth, for example, may contain fibers of wool, of cotton, and of rayon. But in selecting a fabric for a new suit, we are often not concerned as much with the percentage of the different fibers the cloth contains as with its appearance and texture. The utility of the cloth, however, depends more on its constituents than it does upon its eye appeal.

The organs of digestion which prepare food for the needs of the body's tissue utilize the various food constituents rather than handle the types of food we find on the shelves at the market. We possess no separate digestive process for apples as contrasted with potatoes. An apple consists of water, some fruit sugar, traces of protein and fat, some calcium and iron, and several of the vitamins. The digestive organs make these constituents available to the tissues of the body. One food differs from another in the proportions of these various constituents and in the kinds of protein, kinds of fat, and kinds of carbohydrate that each contains. Minerals and vitamins are important, too, and they occur in varying amounts and kinds in the different foods we eat.

From our discussion thus far it may seem that the body is very precise in its need for certain food constituents to play specific roles within the body. If so, we could easily liken the body to an office building under construction. The construction engineer carefully specifies that copper tubing goes in this particular place, a certain grade of steel in another location, and that concrete of a definite grade be used in the footings whereas concrete of a slightly different composition must be used for the walls. The construction engineer, incidentally, carefully guards against substitution for the materials as ordered.

Similarly, maintenance of the human body requires just the right material in each tissue as does the construction of a building. But, fortunately, the individual need not give careful attention, as does the construction engineer, to make sure that each day's food rations contain these exact materials. The chemical processes carried on within the body make it possible to convert certain food constituents into others, as needed. For example, protein primarily builds and repairs tissue. But when the protein intake exceeds the body's present needs, the body converts it chemically into energy-producing food.

Suppose, again, that a person's diet contains more carbohydrate than he needs to provide that day's requirement of energy. The extra carbohydrate is converted to fat, which the body stores. Hence, a person whose diet contains an excess of starch (one of the forms of carbohydrate) may gain weight as his body converts into fat the starch which he eats. A person who wishes to avoid becoming overweight needs to do more than merely avoid the fats he ordinarily eats, for he can become overweight just as readily by eating too much carbohydrate.

The body can also transform stored fat into glucose (blood sugar) whenever the body runs short of energy food.

Suppose a person who ordinarily follows a sedentary way of life decides that on a certain vacation he will climb a mountain. His usual diet provides enough energy food to take care of his limited physical activities. On the day he climbs the mountain, however, his need for energy food may double or, perhaps, triple. Even though he may have had a good breakfast on this day, the unusual activity of his muscles soon uses up the energy food which his breakfast provided. At such times the reserves of fat which the body has stored throughout his tissues provide the additional energy food which the day's activities require.

If this same person continued his mountain climbing for several days without increasing the amount of food eaten, he would lose several pounds of weight because of converting fat stored in his tissues to provide fuel for the greater activity of his muscles.

For the most part, the digestive organs, which prepare food for utilization by the body's tissues, operate automatically. You choose the kinds of food you eat, how much to eat, and when to eat. Beyond this, you simply deliver the food to your mouth and go about your regular activities while the digestive organs do the rest.

While in the mouth, the food mixes with saliva produced by glands located in the walls of the mouth. The saliva not only moistens the food, thus making it simpler to swallow, but contains an enzyme called ptyalin. (Enzymes hasten chemical reactions but do not themselves enter into these reactions.)

Ptyalin begins the digestion of the carbohydrates. Many kinds of carbohydrates exist, but they all consist of molecules composed of just three elements: carbon, hydrogen, and oxygen. The molecules of carbohydrates consist of long chains of carbon atoms with the hydrogen and oxygen atoms fastened to them. As carbohydrates occur in the food, we eat many of the molecules too large to pass through the wall of the small intestine, which absorbs the food the body needs. So, beginning in the mouth, the ptyalin breaks the larger carbohydrate molecules apart to form smaller ones.

You can even taste the difference as this chemical action takes place. While chewing something composed mostly of

How Food Becomes Life

starch (as a soda cracker), you can notice an increasing sweetness as the large molecules of starch break down into simple sugar molecules.

Recognizing that the digestion of carbohydrates begins in the mouth, you should not hasten the process of chewing your food. Allow time enough for the saliva to mix well with the food so that the ptyalin can do its important work.

Once swallowed, food remains in the stomach up to three or four hours. Here it is thoroughly mixed with the gastric juice, which begins the digestion of proteins.

The story of the gastric juice is interesting and remarkable. About thirty-five million tiny glands located in the wall of the stomach produce the gastric juices, one of which is an acid (hydrochloric acid) so strong that it could destroy a piece of ordinary metal, yet the delicate lining of the stomach is normally protected against destruction by this strong acid.

We understand what damage the acid can do to the body's tissues when we observe what takes place in a case of stomach ulcer. In such a situation, a small area of the stomach's lining loses its capacity for protection, and an area of raw tissue (an ulcer) results.

In addition to producing acid, the glands in the walls of the stomach produce an enzyme (pepsin) which, working in collaboration with the acid, begins the digestion of the protein in the food already swallowed.

Proteins, composed of giant molecules, contain atoms of carbon, hydrogen, and oxygen just as in the case of carbohydrates. In addition, however, all protein molecules contain nitrogen atoms and, occasionally, such other atoms as sulfur. The large protein molecules really consist of chains of smaller molecules just as a freight train consists of individual freight cars linked to each other. We call these smaller molecules which compose the large protein molecules amino acids. There are some twenty kinds of these, and the differences between one type of protein and another depend upon the particular kinds of amino acids present.

Before the body's tissues can absorb protein, its large molecules must break up into the component molecules of amino acid. These are small enough that the tissues absorb them as the food passes through the small intestine. From these amino acids, the body builds up its own proteins, as it selects just the right amino acids and unites them to form large molecules.

Proteins constitute the building material out of which the body constructs tissue. Proteins primarily build or replace tissues. Protein is important in the diet of a growing child. Even the adult needs an adequate amount of protein in his diet because of the replacement of worn-out tissue.

A few kinds of amino acids the body cannot build on its own and must depend, therefore, on foods containing them. For this reason, among others, a person needs to eat a variety of foods to make sure he includes all of the necessary nutrients.

After leaving the stomach, the food moves into the small intestine. Here other digestive juices come into play. These have either been produced by glands in the wall of the small intestine or by the pancreas, which discharges its digestive enzymes into the small intestine. Also the liver and gallbladder add bile to the food in the small intestine.

The small intestine completes the breakdown of carbohydrate and protein molecules and also prepares the fat for absorption. In the small intestine the smaller molecules derived from the large molecules of carbohydrate, protein, and fat transfer from the space inside the intestine to the tissues proper, where either the blood or the lymph convey

How Food Becomes Life

them to the various parts of the body where they are needed as energy food or building materials.

In addition to the carbohydrates, proteins, and fats, food substances contain minerals, vitamins, and roughage, all important to the welfare of the body's tissues.

Calcium and phosphorus strengthen the structure of bones and teeth. Iron forms a necessary part of the hemoglobin molecule, which conveys oxygen and carbon dioxide. Sodium, potassium, iodine, and certain trace elements in addition to those named are important to various bodily functions. Vitamins regulate growth and control other of the body's activities. Roughage, consisting mostly of cellulose—an inert material unaffected by the digestive processes—passes to the large intestine and aids in the elimination of the body wastes.

A verse of Scripture states a principle that should guide all serious-minded persons in the matter of eating. Writes King Solomon, "Blessed art thou, O land, when . . . thy princes eat in due season, for strength, and not for drunkenness!" Ecclesiastes 10:17.

King Solomon, wise and wealthy as he was, doubtless used an illustration based on what he saw in his own home. His comment, however, applies to persons at all economic levels. In this context, we do not expect the word *drunkenness* to mean only intoxication by the use of liquor but also the kind of gluttony in which one primarily eats to please his appetite and to encourage self-indulgence.

In recognition of the remarkable manner in which the organs of digestion prepare the food we eat to become a part of the human body, both functionally and structurally, it behooves us, as intelligent persons, to choose our food wisely both in quality and in quantity.

YOUR INCREDIBLE LIVER

7 The liver, the body's largest gland, weighs about four pounds in the adult and performs more than a hundred important functions necessary to health. The lowest five or six ribs protect this soft, fleshy organ located just below the diaphragm, more to the right than to the left.

The ancients believed that passion and desire originated within the liver, and a cowardly person, so they thought, possessed a poorly functioning liver. Even within the present century, many have supposed that constipation indicates something wrong with the liver.

Now we realize that the liver has nothing to do with traits of personality except as a person's state of general health may influence them. Nor does the liver regulate the flow of intestinal contents. Instead, the liver functions as a miniature multipurpose manufacturing plant, producing blood cells in early fetal life, producing body heat, storing and releasing certain substances the body urgently needs, producing bile—an important aid to the body's assimilation of fat—producing important protein substances for the blood plasma, bringing about chemical transformations in many substances which the body requires for energy or tissue-building, preparing leftover or expended substances for elimination, and rendering certain toxins less noxious.

All of the liver's functioning cells lie close to flowing

blood. In fact, blood bathes practically every liver cell on at least one of its surfaces. This is important, because the liver performs its chemical magic on substances brought in and carried away by the blood. It is no surprise, then, that the liver has a double blood supply—one consisting of fresh blood coming quite directly from the lungs and heart, and the other of blood which detours through the liver after it leaves the intestine for the heart. The blood from this second source carries the food substances which it just absorbed through the wall of the intestine.

The early development of an unborn child demands a large production of blood. During this time the liver manufactures blood cells, which function it transfers to the bone marrow and the lymphoid organs by the time of birth. Specialized liver cells lying between its ordinary gland cells and the stream of flowing blood, take on this temporary task, and once the job of producing blood cells ends, these special cells seem to disappear, only to reappear and resume work at any time in later life should the supply of blood cells drop to very short supply.

The gland cells which compose the bulk of the organ carry on the remaining functions. The liver's ability to perform many tasks becomes the more remarkable when we notice its nearly homogeneous structure. No special regions in the liver perform its sundry functions, and all gland cells of the liver look alike. It seems that the same cells capably switch jobs as the occasion requires.

An important incidental function of the liver is to produce heat. Body heat develops from the union of oxygen with energy-producing food materials. This is the same means of heat production we find in the house—a union of oxygen with fuel, but in the body the process occurs relatively slowly and under controlled conditions.

Admittedly, the muscles, when active, produce more body heat than the liver. However, when the muscles rest, the liver produces the greatest amount of heat. Hence, some call it the body's "internal furnace."

Despite the liver's many activities, it also serves as a storehouse. Although the blood passing through the liver always keeps moving, the amount of blood contained within the organ varies from time to time. When it distends with blood, it serves, to this extent, as a reservoir.

The liver also amasses within its cells sufficient supplies of the fundamental food elements to liberate into the blood as needed to meet the body's moment-by-moment requirements. For example, the liver reserves a sufficient amount of carbohydrate (the body's prime energy food) to provide up to five hundred calories of food energy. True, this is only about one fifth of the average daily energy requirement; but it serves as a working capital always available on short notice to fill the need while the body mobilizes other less accessible stores of energy food.

In a similar manner the liver stocks modest amounts of amino acids—the building blocks for the various proteins—so that the body can quickly assemble just the particular kind of protein that may be momentarily in short supply. The liver also stores a certain amount of fat, even though most of the body's reserve supply is located elsewhere.

The liver contains deposits of vitamins A, D, and B complex, holding these ready for utilization by other tissues.

The liver husbands the body's precious store of iron. Even though iron abounds in the mineral deposits of the world about us, the human body receives relatively little of this vital material needed in the formation of hemoglobin. The hemoglobin in the red blood cells enables them to carry oxygen from the lungs to the various tissues and carbon dioxide from the tissues back to the lungs. Should the amount of iron in the body fall below the desirable minimum, a form of anemia develops in which the whole system suffers because the blood cannot transport oxygen in adequate quantity.

Even though the food eaten may include considerable iron, there is a limit to the amount the body's tissues can absorb each day. Thus the body's stores of iron build up slowly. Since the body eliminates a certain amount daily, the balance between the intake and outgo of iron becomes an important item, and here the liver comes into the picture by helping to maintain a favorable balance between iron intake and expenditure. To understand this, we must mention briefly the life cycle of the red blood cells.

Five million red blood cells float in a cubic millimeter of blood. This means that the body of an average-sized adult contains something like 25 trillion red blood cells at any one time. A red blood cell lives about 120 days, which means that the average adult body destroys and replaces about 150 million red blood cells every minute. Most of this destruction of the worn-out red blood cells occurs in the spleen. As it destroys these cells, their chemical constituents seep into the blood, and as the blood comes through the liver, it retains the iron, for the most part, which it then makes available to the blood-forming tissues as needed for the production of new red blood cells.

The liver also produces bile, and in this connection the liver comes closest to qualifying as a typical gland, for it includes a system of tiny canals which carry away the bile secreted by the individual gland cells. These tiny canals unite to form ducts, which in turn unite to form a single passageway which carries the bile into the small intestine. The liver makes between a pint and a quart of bile a day.

The bile salts emulsify the fat contained in the food passing through the intestine and thus make it available for the chemical changes that take place before the blood absorbs it. Not only does the bile aid in the digestion of fats, but it also makes possible the absorption from the intestine of the so-called fat-soluble vitamins—vitamins A, D, E, and K—as well as the carotene out of which vitamin A is produced.

In addition to salts, bile also contains pigments which represent some of the end products in the destruction of hemoglobin. These account for the characteristic yellow color (jaundice) of the skin and eyes when something obstructs the flow of bile.

By now you can understand the importance of the liver. Perhaps you are saying to yourself, "It seems that the liver performs most of the important functions of the body."

And it is true; the liver plays a vital role. The spleen, for example, can be removed, and still the usual processes of life continue quite normally. The gallbladder and the appendix are easily expendable. But not so with the liver. When disease affects the liver, the entire body suffers; and if the disease progresses, life itself is endangered.

We might liken the liver to the maintenance department of a great institution. It takes care of the routine matters. Its work is unglamorous but important. The maintenance department makes no major decisions; the administrative officers make those. The maintenance department does not buy or sell; rather, it is responsible for maintaining the internal operations of the institution.

So with the liver in its relation to the rest of the body. The liver performs the important functions involved in maintaining the body's internal economy.

It produces certain proteins for the blood plasma—the

chief being albumin, which cushions the blood against sudden changes in relative acidity and provides just the right osmotic pressure to prevent excessive loss of water from the bloodstream. Other proteins produced by the liver and which circulate in the blood plasma have to do with the clotting of blood: fibrinogen and prothrombin.

We can describe the liver's chemical wizardry thus: The liver takes the materials brought to it by the blood, either from the digestive organs or from remote parts of the body, and custom-builds from these the particular kinds of chemical substances required by the body. Take fat as one example. The chemical composition of human fat differs from that contained in tissues of other animals. When we eat fat of animal origin, the liver must dismember the molecules of this fat and rebuild them to form the exact kind of fat which the human body can store or utilize.

While we speak of the liver's ability to transform chemical substances, we should emphasize that the most marvelous manifestation of this capacity is the conversion of fats into carbohydrates, carbohydrates into fats, or proteins into carbohydrates, as the needs of the body may require. Consider for a moment what a hardship you would suffer if you had to select your food each day to provide the exact proportions of the three fundamental food constituents your body needs. But instead of worrying about these details, you rest assured that your liver will do the job, just so long as you select a wholesome and reasonably adequate diet.

The liver also dismembers the molecules of certain worn-out substances that the body's tissues discard. In this connection the liver primarily disposes of discarded protein. Protein is unique among the basic food substances because it contains nitrogen and because its molecules are so large the normal kidney cannot eliminate them. The liver therefore breaks down these large protein molecules and salvages the parts it can convert to other uses. The residue containing nitrogen forms urea, a substance consisting of small molecules, which the kidneys readily eliminate.

Finally, in our consideration of the functions of the liver we come to its influence in making certain poisonous or harmful substances less deleterious. The exact means by which the liver accomplishes this varies from substance to substance; but, in principle, the chemical agents available in the liver either disrupt the molecules of the noxious substance, rendering them harmless, or they effect chemical combinations which have similar results. The body then eliminates the final products of these harmful substances.

Considering the many things the liver does to further the welfare of the body, one can easily understand that disease of the liver, in any form, poses a serious threat to health and even to life itself.

Under normal circumstances, however, the liver possesses a wide margin of safety, since it contains much more tissue than it actually needs to carry on its usual functions. All that a person can do to protect and ensure the continued health of his liver is to follow a temperate, health-promoting way of life.

Humble yet indispensable, the liver carries on its important functions twenty-four hours a day. So with many of the affairs of life—those activities which appear routine and which attract relatively little attention frequently are the most important in furthering the welfare, comfort, and progress of the persons concerned.

KEEPING WHAT YOU NEED

In the San Bernardino Mountains of southern California, a populous resort community has developed on the shores of Big Bear Lake—an artificial lake created years ago when the Big Bear Water Company constructed a dam across a narrow part of the canyon that formerly served as an outlet for Big Bear Valley. The impounded water formed a lake seven miles long and nearly a mile wide in some places.

In recent years the water level in the lake has lowered because of the increased use of lake water for irrigation in the valley below. Those who live at Big Bear Lake have stated in court that the interests of their community require that the use of the water for irrigation must stop. The water company contends, "We built the dam in the first place, therefore we should be free to use the water as we please." But the community representatives retort, "We need the water, and we want to keep it."

The circumstance at Big Bear Lake illustrates some of the relationships within the human body. Here it is not a simple matter of water rights, even though the regulation of the amount of water which the body contains is important. It is a matter of the body's ability to preserve for its own use all needed materials.

The body requires many different substances in order to

function normally. The food and drink and air that we take in supply these. Many of the needed materials occur in such abundance that the excesses must be thrown away along with the body's waste products.

It would seem that in this elimination of useless substances the danger exists that some essential materials may also be lost. But a healthy body has the marvelous ability to keep just what it needs. The kidneys constitute the organs that largely determine what shall stay and what shall go. Take the body's need for water as a first example. A little more than half of the body's weight is due to the water it contains—water in the blood plasma and other body fluids, water within the body's cells, and water in the spaces between the cells.

The body constantly loses water by several means. Exhaled air carries water vapor with it. This accounts for the loss of almost a pint of water a day. Between one and two pints are lost from the skin each day even during moderate external temperatures. During physical activity or high outside temperatures a person can perspire as much as two quarts an hour. An average of three pints of water goes out in the urine. The waste material expelled from the rectum contains a small amount of water. Notice the tremendous variation in the amount of water eliminated, the exact amount depending on conditions of the moment.

The body has three sources of water: (1) the water which a person drinks (an average of three pints a day), (2) the water contained in the food eaten (averaging one quart a day), and (3) the water produced within the tissues as the result of processes of oxidation (averaging about one-half pint a day).

Water—the universal solvent—is vital to the body's economy. All the chemical processes of the body involve

Keeping What You Need

molecules of substances that are dissolved in water or else involve the water molecules themselves.

The exact amount of water which the body contains at any given time obviously depends upon the balance between the intake and the outgo. Loss of water from the body occurs easily and may occur very suddenly in excessive perspiration due to overheating, in vomiting, in diarrhea, and in hemorrhage. The body can adjust to small variations in the amount of water, and when the supply runs low, thirst prompts the person to take a drink. When the amount of water accumulates beyond what is needed, the kidneys eliminate the excess. If, in extreme conditions, the loss of water amounts to 10 percent of the body weight, the situation is serious. If the loss approaches 20 percent of the body weight, death could ensue.

As the blood passes through the kidneys, a certain portion diverts from the main stream and moves more slowly through the kidneys. Some of the blood plasma seeps into the kidney tubules for a sorting out of its chemical constituents. This dilute solution which contains no protein (nor does it carry any blood cells) is called the "filtrate." During a twenty-four-hour period the total filtrate amounts, on the average, to about 180 quarts—many times the volume of the body's blood plasma. Obviously, then, the kidneys overhaul the fluid portion of the blood many times each day as they check and alter appropriately.

As the filtrate moves slowly through the kidneys' tubules, specialized cells recover most of the water and dissolved substances which are still useful in the body and return them to the main bloodstream. At first mention, this seems like a tedious process. We must realize, however, that water "flushes out" the waste products and the excesses of other substances contained in the blood plasma.

About 85 percent of the water contained in the filtrate automatically returns to the bloodstream. What happens to the remaining 15 percent depends upon the momentary water balance within the body's tissues. If the tissues and blood contain plenty of water, the kidneys allow a surplus of water to pass into the bladder to become part of the urine. If water is in short supply throughout the body, most of the water in this remaining portion of the filtrate recirculates so that in extreme cases only about one pint of water a day goes to the bladder for elimination.

This final control is influenced by a tiny bit of tissue which serves as a monitor to measure the concentration of the blood. When the blood becomes too concentrated, the pituitary gland secretes a chemical substance known as vasopressin. The blood carries it to the kidneys, where it brings about the return of as much water as possible to the blood.

Here we have a mechanism that reminds us of what takes place inside a chemical manufacturing plant. As we contemplate it, we can almost visualize a group of legendary dwarfs residing within the kidneys, waiting for signals from the pituitary gland. At the signal, they either close or open the figurative valves which regulate the amount of water that enters the urine.

The kidneys have much else to do in addition to controlling the body's water balance. They not only permit the elimination of useless materials, but prevent the loss of valuable substances. It is as though they put the molecules dissolved in the kidney filtrate through a chute, saying to this molecule, "You go straight ahead," and to another, "You take the next exit to your left."

It is ridiculous to assume that the infinitely large number of molecules contained in the filtrate run single file past

Keeping What You Need

a solitary checkpoint. The kidneys handle their task very much as ticket checkers at a large public performance handle the crowd of people as they enter. In such a case there are many ticket checkers and many turnstiles so that only a fraction of the large crowd enters by any one gate.

In a similar manner each kidney contains about one million identical structures, called nephrons, each of which deals with one-millionth part of the filtrate. So, with the two million nephrons operating in the two kidneys, the work goes forward with dispatch.

A nephron consists of two parts—a renal corpuscle and a tubule. Each renal corpuscle consists of a tuft of blood capillaries surrounded by a delicate capsule, shaped like a funnel, which collects the fluid as it seeps out of the blood capillaries and pours it into the tubule. Tiny openings in the walls of these capillaries allow water and other substances to pass through. The openings are purposely so small that the giant protein molecules contained in blood plasma and the blood cells cannot squeeze through.

In addition to water, the substances which the kidney tubules reclaim include glucose (blood sugar), carbonates, sodium, chlorides, potassium, most of the phosphate, most of the calcium, and a small part of the sulphate contained in the filtrate. Also, the tubules reabsorb part of the urea (the principal nitrogenous waste product) only to eliminate it at a later time when it passes through the kidney again.

We should underscore the importance of recovering glucose from the filtrate. Glucose constitutes the body's tissue fuel. By uniting with oxygen in the various cells of the body it provides the energy with which they carry on their activities.

In a healthy person under normal circumstances, all of the glucose returns to the circulating blood plasma for use throughout the body. Under some circumstances, however, the amount of glucose in the circulating blood plasma becomes too great. In such cases, tubules reabsorb only part of the glucose, allowing the excess to pass into the urine.

When a person has eaten too much candy or when for any reason a large allotment of sugar overwhelms the liver, the concentration of glucose in the blood reaches dangerous proportions. This is one of the circumstances under which the kidneys allow the excess of glucose to pass into the urine. A similar situation occurs in diabetes in which the tissues fail to use glucose in the normal manner. Then glucose "spills over" into the urine. This explains why the detection of glucose (sugar) in the urine (by a simple laboratory test) is an important aid in diagnosing diabetes or in the handling of such a case under treatment.

Protein, a substance for which the body has urgent need, does not become a part of the filtrate. We mentioned, in describing the minute structure of the kidney, that the openings in the walls of the capillary tufts inside the renal corpuscle are so small that they do not permit the large molecules of protein to pass through. For this reason protein never appears, normally, in the kidney filtrate and hence not in the urine.

During certain forms of illness, however, protein does appear in the urine. When present, it indicates that the capillaries in the renal corpuscles are damaged, thus allowing the protein to seep through and become lost to the body. This sort of damage occurs in many cases of acute poisoning, as a complication of some of the infectious diseases, or in degenerative conditions of the kidney.

Remarkably, the complicated, vital functions of the human body proceed automatically. How fortunate that we do not have to give conscious attention to the amount of

Keeping What You Need

water we should eliminate from the body on a given day! Neither do we have to count the sodium ions or the phosphate radicals or the calcium atoms to be sure there are a sufficient number in the body.

Normal eating and drinking supplies the body with a sufficient amount of the materials which it needs, and the kidneys—credit to them—take care of the details of maintaining the right amounts of these substances in the body. Of course, they cooperate with other organs, but the kidneys have the final say on which molecule stays and which one goes.

YOUR BODY'S SUBCONTRACTORS

In a building operation or a manufacturing enterprise usually the owners employ a responsible contractor to produce the building or the manufactured goods as specified by the owner. But likely as not this master contractor does not want to concern himself with many of the small operations the job involves. So he hires subcontractors who specialize in various fields.

A responsible subcontractor relieves the master contractor of worry, but as a precaution, to make sure that the subcontractors do their work well and particularly to time their activities so that they fit well into the general operation, he may employ a super-subcontractor who double-checks and controls the work of the smaller operators.

In the workings of the human body we find a similar organization. The brain is the master contractor for the body's functions. It controls all, and the many nervous connections which extend throughout the body keep it informed of the activities and conditions in all areas. Also controlling nerve impulses which originate in the brain flash to many parts of the body so that these parts can cooperate in whatever the brain determines to do.

But the brain has more important things to do than to make sure each cell carries on its work as it should. So it sublets to the endocrine glands the jobs of determining how

YOUR BODY'S SUBCONTRACTORS

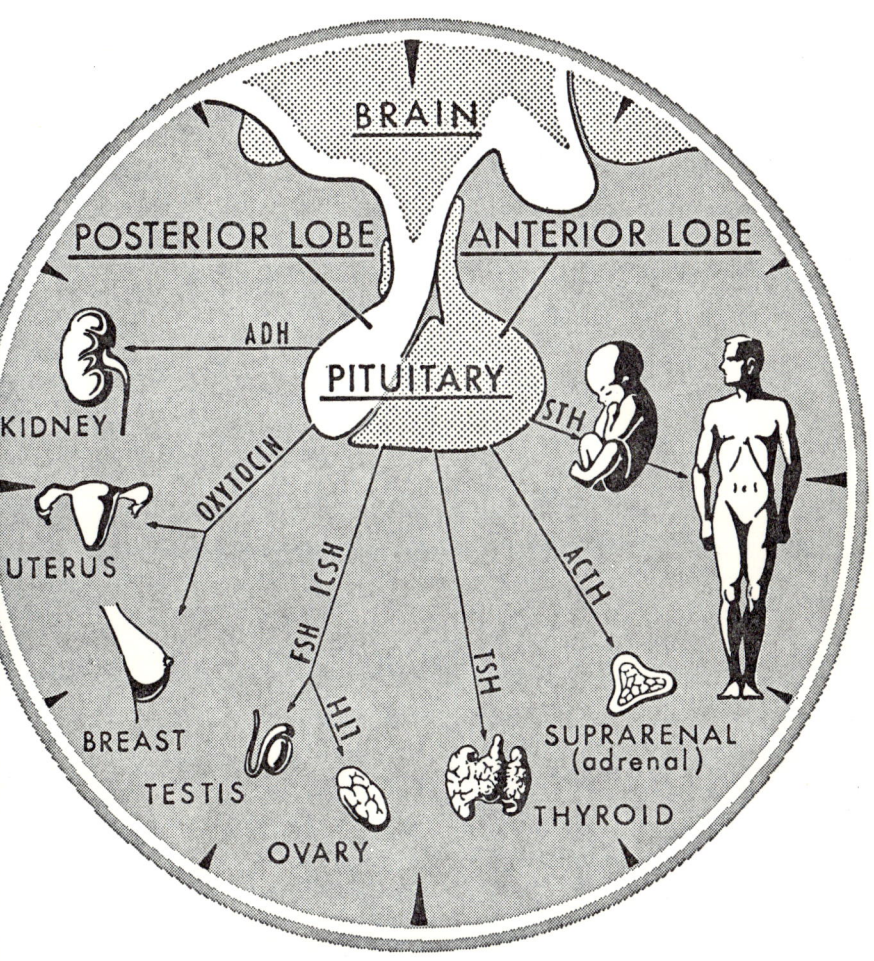

fast the tissues shall grow, of making sure that enough blood sugar meets the demands of an emergency, of controlling the cyclic function of the sex organs, of making sure that the body retains the proper amount of water, and of regulating many other minor functions. These "subcontractors," which control many details of the body's functions, include the pituitary gland, the adrenal glands, the thyroid gland, the parathyroid glands, a portion of the pancreas, and the sex glands.

And, sure enough, just as a large manufacturing operation needs a super-subcontractor, so, in the organization of the endocrine glands, the pituitary gland fulfills this function. As indicated in the drawing, the pituitary gland calls the signals for most of the endocrine glands as well as for some of the body's other tissues.

In an endocrine gland highly specialized cells produce a chemical not needed by the cells which produce it but which is important to the welfare of other parts of the body; the blood carries this chemical to the areas it influences. Notice that blood transports the products of the endocrine glands. In this respect the endocrine glands differ from such glands as those which produce saliva or the glands in the skin which manufacture sweat or the glands in the wall of the stomach which secrete gastric juice. These other glands produce fluids which pour out through a duct onto the body's surface or into the spaces within the body's organs. Because the endocrine glands have no ducts, but deliver their secretions directly to the blood, they are sometimes called ductless glands.

Scientists call the chemical substances which the endocrine glands produce hormones. Interestingly, even though the blood carries a given hormone to all parts of the body, this hormone affects only those cells which it is intended to influence. For instance, the pituitary gland produces a unique hormone with the specific function of inducing greater activity in the thyroid gland. The blood transports this thyroid-stimulating hormone produced by the pituitary gland to the hands, the feet, and all other parts of the body.

Your Body's Subcontractors

But the only cells in the whole body which respond are the thyroid cells.

Inasmuch as the pituitary gland controls most of the other endocrine glands, we can more easily understand how the endocrine glands operate by considering one at a time the hormones which the pituitary gland produces and then tracing these to the "target organs," or tissues, which they influence. The reader will find this relatively simple if he follows the symbols contained in the accompanying drawing beginning with STH, in the three o'clock position, and continuing in a clockwise direction.

The drawing clearly shows the two parts of the pituitary gland—an anterior lobe and a posterior lobe. The anterior lobe secretes six hormones and the posterior lobe two.

STH (somatotropic hormone) promotes body growth. It affects all those organs and tissues of the body which influence the process of growth. When the pituitary gland produces too little of this hormone, a child does not grow as much as he should, and he becomes a dwarf. This type of midget is well proportioned and normally intelligent; he is merely built on a smaller scale than the average person. When STH exceeds normal, a child will grow to larger proportions than normal and will become a giant. Thus the eight-foot man in the circus has a pituitary which produced too much STH throughout his childhood.

The second hormone produced by the pituitary gland is ACTH (adrenocorticotropic hormone), which stimulates the cortex of the adrenal glands to produce certain corticoids. The corticoids manufactured here convert protein and fat to carbohydrate and also regulate the proportions of potassium and sodium in the body.

The cortex of the adrenal also secretes a hormone called androgen, which, as it circulates through the body, has a masculinizing influence. Different from the hormones of the male sex glands, it has a similar effect. When this hormone is produced excessively in a woman, she develops hair on her face and other evidences of masculinity.

The third hormone made by the pituitary gland is TSH (the thyroid-stimulating hormone). When TSH reaches the thyroid gland, it stimulates the production of the iodine-containing hormones, which have several important effects in the body. First, they increase the rate of the body's metabolism (the rate at which it uses oxygen). Second, they improve a person's ability to think. When a person does not have enough of these hormones, he is sluggish, but when he has too much, he grows restless and irritable. The hormones manufactured by the thyroid gland also stimulate normal growth and skeletal development in a child.

The remaining three of the hormones produced by the anterior lobe of the pituitary gland are classed together as the gonadotropic hormones because they influence the activities of the sex organs. FSH aids the manufacture of sex cells by the ovaries in a woman and by the testes in a man.

ICSH (interstitial-cell-stimulating hormone) encourages the production of the sex hormones. In a woman, ICSH increases the output of estrogen, which, in a girl, stimulates the development of the feminine characteristics and, in a woman, helps regulate the functions of the reproductive organs. ICSH in a man stimulates the production, in the testes, of testosterone which, in an adolescent boy, stimulates the development of the secondary male characteristics and, in a man, maintains these same characteristics.

LTH, the third gonadotropic hormone, affects only women. It stimulates certain tissues of the ovary to make progesterone, which prepares the lining of the uterus (womb) once a month to receive and nourish the newly

Your Body's Subcontractors

formed cells of a future child, if pregnancy occurs.

The posterior lobe of the pituitary produces two important hormones, oxytocin and ADH. Oxytocin has no demonstrable effect in a man's body, but in a woman it helps the uterus and breasts fulfill their normal functions at the time of childbearing.

ADH (antidiuretic hormone) acts on the cells in the tubules of the kidneys in such a way as to decrease the amount of water that passes through the kidneys and becomes part of the urine. Thus, it plays a very important role in maintaining the proper balance of water within the body.

Having now completed our description of the hormones produced by the pituitary gland, we have two other endocrine organs to mention which the pituitary seems not to control. First, the parathyroid glands, small organs, usually four located in the neck, some on the right and some on the left, on the back surfaces of the lateral lobes of the thyroid gland, produce a hormone called parathormone, which controls the amounts of calcium and phosphate circulating in the blood. The parathyroid glands are essential to life, for when parathormone is not present in the blood, the amount of calcium in the blood falls to such low levels as to cause the death of the individual.

The pancreas is a fairly large organ located in the abdominal cavity, close to the stomach. Scattered throughout the substance of the pancreas are small islands of tissue known as islets of Langerhans. These islands of tissue principally produce insulin, which controls the use of glucose (blood sugar) by the body's various tissues. Adequate amounts of insulin enable the tissues to use glucose freely. A lowered amount of insulin deprives the tissues of part of this energy fuel, which they need, and large amounts of sugar accumulate in the blood. The usual kind of diabetes results from an insufficient amount of insulin.

Now that we have completed our description of the endocrine glands—the body's subcontractors—we should pause a moment to reflect on the marvelous design and function of the human body. The brain supervises the body's activities. In the interests of efficiency, however, many functions take place automatically or under the precise control of subcontractors, such as the endocrine glands. This allows opportunity for the brain to consider matters of cultural, intellectual, and spiritual import. Our brains can even appreciate the remarkable manner in which the Creator designed our bodies.

HOW LIFE BEGINS

Some treatises on the human body omit the reproductive organs from the list of "vital organs" by labeling them "nonessential to life." This means that a person's heart will still beat, his digestive organs will still function, he can still take one breath after another, and his brain will still think, even in the absence of the reproductive organs.

But this casual attitude which assigns the reproductive organs to an accessory role does not do them justice. In the first place, without the reproductive organs a person would be neither a man nor a woman. Both in body build and temperament he would continue his childlikeness even in adulthood.

Such a person would lack the strength and aggressiveness which characterize a man. His shoulders would not be broad, he would have no beard, his voice would be mild and shallow.

Just as the tissues of a man respond to the male hormone produced by his sex organs, so in the female body the hormone which is produced by the ovaries and circulates in the blood triggers and maintains the change from girlhood to womanhood. This hormone broadens the hips and makes the voice melodious.

Thus we recognize that the reproductive organs have more than one function to perform. The first function, already mentioned, that of producing the sex hormones, begins in the early teens, at the time of transition from childhood to young adulthood. The sex hormones, produced in the testes of a boy and in the ovaries of a girl, work in conjunction with the hormones produced by the pituitary gland to effect those changes in the body and in the personality which prepare the individual to fulfill his place in life, either as a man or as a woman.

How fortunate that in nature's sequence of events men and women prepare for their future roles of fatherhood and motherhood by first confirming membership in their respective sexes! This allows time for a prospective father to establish himself as a man and to adjust to his masculine status so that later when he becomes a husband and father he will make wise use of his natural qualities of leadership, courage, and fortitude. Similarly, it is good for a girl to have a little time during her middle and late teens to learn how to identify herself adequately as a woman and thus, as she joins in establishing a family, bring to her husband and her children the benefits of her affection, tenderness, and unselfishness.

The second function of the reproductive organs is better known—that of passing on the "spark of life" from one generation to the next. The sex cells produced by these organs, once they unite, initiate the life of a member of the new generation.

The reproductive organs preserve the hereditary legacies of generations and make them available as an endowment to children yet unborn. These organs permit a biological blending of the genetic contribution of a husband with that of his wife so that as they enter parenthood their child duplicates neither, but is an amalgamation of both. The female organs foster, protect, and nourish this newly formed life for the critical first nine months of its existence, until it grows large enough and strong enough to become a member of the family.

When a girl enters young womanhood, her relatively undeveloped reproductive organs respond to the hormones now produced by the pituitary gland and begin functioning according to the pattern of womanhood.

The two ovaries, located low in the abdominal cavity, one on the right and one on the left, produce both the female hormone and sex cells. Almond-shaped, they measure about one and one-half inches long, a little less than one inch wide, and about one third of an inch thick.

At the time sex life begins, the surface layer of each ovary contains several thousand immature female sex cells. From then on a few of these respond each month to the stimulating influence of the hormones which control the monthly sexual cycle in a woman's body. As they respond, these few cells grow rapidly and seem to compete with each other for the chance that month to unite with a male sex cell and thus begin a new life.

Sometimes a sex cell from the right ovary and sometimes one from the left liberates itself during a particular month. If the left ovary produces the "favored" cell this month, next month it will probably be the right ovary. In the rare case in which one ovary has become diseased or has been surgically removed, the remaining ovary produces the favored cell month by month. In rare instances, two female sex cells are liberated during the same month, and if both of these cells should unite with male sex cells, twins result.

The ovaries are located on either side of the uterus—a pear-shaped organ, about three inches long and two inches wide, situated in the center of the lower part of the abdominal cavity, large end up. It is hollow with thick walls, and on each side, right and left, a slender tube, the oviduct, connects with the cavity inside the uterus and extends away from the uterus, on its own side, into the abdominal cavity near the ovary.

Now back to our story of what happens when a female sex cell is liberated, one a month, from one of the ovaries. The open end of the oviduct moves close to the ovary, and the sex cell enters the oviduct for a four-day trip toward the cavity of the uterus. Microscopic fingerlike structures located inside the oviduct gently move it on its way.

This event, in which a female sex cell passes through one of the oviducts toward the cavity of the uterus, occurs at about the middle of a woman's menstrual cycle—about two weeks, in the usual case, after the beginning of the previous menstruation.

In the event no male sex cell unites with the female sex cell as it passes through the oviduct this particular month, the female cell moves on into the cavity of the uterus and eventually perishes.

In order to understand the events that occur when it becomes possible for the female sex cell to unite with its male counterpart, we must turn our attention, next, to what takes place within a man's reproductive organs.

The two testes (singular, testis) are the essential organs of the male. Located in a fleshy sac, called the scrotum, which hangs from the lower part of the abdomen, between the upper parts of the thighs, the testes are oval-shaped organs about one and one-half inches long, one inch wide, and a little more than one inch thick, from front to back.

The number of sex cells produced by the testes far exceed the number of sex cells produced by the ovaries. As many as one hundred million male sex cells per day flow from the testes into a complicated system of tubules which

How Life Begins

eventually open into single tubes that lead upward into the abdominal cavity to the prostate gland located just beneath the bladder.

The male sex cells have a different appearance, under the microscope, from that of the female sex cells. A female sex cell is spherical, whereas a male sex cell is slender and tapers toward one end to form a so-called tail process, which can thrash from side to side and thus propel the cell.

The testes quite continuously produce the male sex cells; these accumulate in the nearby tubules, which serve as a reservoir. Within them the male sex cells come into contact with an acid fluid secretion which reduces their activity.

At the time the male sex cells discharge from a man's body, tiny muscles within the walls of the tubes containing these cells contract, forcing the sex cells on their way. At the same time glands located along the route of travel discharge their alkaline secretions, which activate the resting male sex cells so that they can propel themselves.

During sexual intercourse, the husband ejaculates his male sex cells into the vagina of the wife. The vagina connects with the lower part of the uterus, and through the uterus, with the oviducts. The male sex cells move about one inch in eight minutes and thus, in a relatively short time, find their way from the vagina into the cavity of the uterus and out into the oviducts on the right and on the left.

If, at such a time, the lone female sex cell happens through the oviduct, it unites with one of the available male cells. This union of sex cells constitutes conception and serves as the means of transmitting life from this generation to the next.

The combined cell, formed by the union of a male and a female sex cell and from which all of the cells which form the body of a new individual will develop, is called the zygote. It continues traveling through the remainder of the oviduct until it reaches the cavity of the uterus, where it attaches to the delicate membrane lining the uterus and by a marvelous sequence of events receives its nourishment here throughout the nine months of growth prior to childbirth.

As we read about the events connected with conception, we realize the complicated adaptations in the bodies of both a mother and a father that make parenthood possible and marvel at the precision of the Creator's handiwork. We begin to understand why the same God who provides for the development and care of an unborn child continues His interest in its welfare throughout its lifetime.

TEMPERATURE CONTROL

In these days of machines, gadgets, and push buttons, the word *automatic* has become common. We speak of an automatic heating and cooling system for the home, an automatic clothes washer, and an automatic sprinkling system. Away from home we have automatic elevators, automatic tollgates, and automatic vending machines. Modern manufacturing plants have so automated that some people fear that machines will deprive workmen of jobs they need.

But with all our pride in recent inventions, let us not assume that automation has developed only in the past few years. The human body is filled with automatic control devices which care for its many functions and leave the brain free for unrelated activities. And the Creator built in these automatic controls from the first.

Take, for example, the body's means of controlling its own temperature, which must remain very close to 98.6 degrees Fahrenheit. All body tissues produce heat by the process of oxidation, in which oxygen combines with food substances inside the cells. Muscle tissue generates the greatest amount of heat, particularly when active. When a person strenuously exercises, his muscles produce twenty times as much heat as is produced by the combined activity of all other tissues.

The blood, as it flows from place to place throughout

the body, equalizes the temperature of all tissues. If one part of the body produces more heat than another, the blood flowing through this part absorbs the extra heat, carries it away, and shares it with all other parts.

When the total heat produced in the body exceeds the total heat dissipated, body temperature rises. Contrariwise, when the total heat produced falls below that being lost, body temperature decreases.

Heat dissipates by the evaporation of moisture on the skin, by the conduction of heat to the air or objects in contact, such as cold bath water, and by radiation from the body. Heat increases, not only through the production of heat by the body's tissues, but by hot food and drink or by skin contact with hot air, hot water, or heated objects.

When the body loses more heat than it gains, it generates sensations of cold, which are carried to the conscious centers of the brain so that the person says to himself, "I feel cold." He can then put on more clothing, step into the sunshine, or move into a warmer room. But by this time some of his body's automatic controls have already gone into operation. Shivering, one of the reflex responses to cold, involves the action of muscles, and muscles, when active, produce heat. Thus when a person becomes cold, his production of body heat steps up.

When body heat accumulates, there develops an automatic response of sweating. The evaporation of moisture from the skin has a cooling effect. Here again the sensations of changing body temperature are carried to the brain's conscious centers. It is likely, however, that the person is already sweating by the time he feels too warm.

The use of muscles as in shivering or in exercise is not the only way to stoke the body's heat-producing mechanism. Emotions of fright and of anger similarly influence the auto-

Temperature Control

matic nervous system to raise body temperature.

In response to distressing emotions, two chemical substances flow into the body's tissues—norepinephrine and epinephrine, which increase the rates of oxidation in all the body's cells. Their effect is so pronounced that they can raise the total heat production in the body as much as 200 percent.

The production of body heat represents an investment of the body's resources, for it takes energy to produce heat. Understandably, then, the body's automatic controls will act to prevent its unnecessary loss.

Just as the householder closes the doors and windows of his house when the outside temperature drops, so the body curtails the supply of blood to the skin to conserve heat. A reflex contraction of the smooth muscle in the walls of the blood vessels carrying blood to the skin constricts these vessels so that they naturally carry less blood, and this reduces the loss of heat from the skin. You have noticed, no doubt, that when your arms are cold, the skin appears pale because less blood flows through the skin than during times of warm outside temperature.

The automatic control goes beyond this, however, in the attempt to conserve body heat. The arteries that supply the fingers and toes have certain bypass branches which ordinarily remain in disuse. These make a direct connection between the arteries to which they belong and the nearby veins. When the body urgently needs to conserve heat, the muscles in the walls of these bypass branches relax so that blood can pass through, moving directly from the parent artery to a corresponding vein, without passing through the capillary network in the more remote parts of the fingers and toes. These bypass connections, called arteriovenous anastomoses, carry the blood from the arterial circuit to the venous circuit while it is still warm. In the meantime, of course, the fingers and toes may become very cold, but they drop in temperature in the interest of maintaining a desirable temperature for the internal organs.

Now that we have considered several of the means by which the body produces and dissipates heat, we should notice how it regulates these. The thyroid gland primarily influences the rate of oxidation that takes place in all the body's cells. This gland produces an interesting hormone, thyroxine, which the chemist describes as a combination of tyrosine, one of the amino acids, with iodine. When the thyroid produces large amounts of thyroxine, all of the processes of oxidation throughout the body proceed at a relatively rapid rate. When production decreases, the oxidative processes proceed at a slow average rate.

The momentary circumstances that require the production of more heat can still affect the processes of oxidation in the cells in a manner that increases heat production for the time being. Similarly, if too much heat is generated, the usual means of dissipating heat goes into effect whether or not the thyroid is producing large amounts of thyroxine.

The thyroid regulates heat production seasonally rather than momentarily. The thyroid of a person who lives where winters are cold is more active during the winter season than during the summer.

The person with an active thyroid gland perspires more readily than other people. His body temperature may even run a little higher than normal. He feels warm when other people put on their jackets.

On the other hand, the person with an underactive thyroid gland may wear his coat when others are in shirt sleeves, for the cells throughout his body do not produce as much heat, on the average, as they normally should.

We have already mentioned the means the body uses to

Temperature Control

produce heat. Also, we have listed ways that it can transfer heat when its temperature begins to rise. But these means of increasing or decreasing the body's temperature must be under one unified control.

In the old-fashioned dwelling house, the control of temperature was relatively simple. When the house became too warm, someone turned down the furnace and, if necessary, opened the windows. When the house became too cool, he did just the opposite—closed the windows and turned up the furnace.

In a modern home, the thermostat automatically controls all of this adjustment of heat production and heat dissipation. When the temperature rises too high, a piece of metal inside the thermostat expands just enough to break an electric circuit and turn down the furnace, or another turns on the cooler, or both. If the house becomes too cool, the thermostat turns off the cooler, and turns on the furnace when necessary.

In the human brain the hypothalamus serves as a thermostat and functions automatically without intruding into conscious thought.

The temperature of the blood as it flows through the hypothalamus influences its temperature control center. Blood which becomes a little too warm activates reflex circuits which have two effects. First, they inhibit oxidation in the cells throughout the body. This parallels turning down the fire in the furnace. Second, if the temperature of the blood passing through the hypothalamus remains too warm, the reflexes stimulate the sweat glands to pour out their secretion on the skin surface. As already mentioned, this cools by evaporation.

If the blood flowing through the hypothalamus becomes even a fraction of a degree cooler than normal, it activates devices for producing and conserving heat. The small vessels in the skin constrict to reduce the amount of heat dissipated from the skin surface. Oxidation increases throughout the body to implement heat production. In more urgent cases, it stimulates the muscles to the point of shivering.

In some conditions of illness a higher body temperature enhances the body's defense mechanisms. Chemical processes proceed more rapidly at higher temperatures, and faster chemical reactions in the body may improve its ability to deal with disease and bring about healing. It may also be, in cases of illness caused by bacteria and viruses, that higher than usual temperatures impede bacterial or viral growth.

At any rate, the thermostat setting automatically and temporarily adjusts upward. Such a general rise in body temperature we call fever.

The human body is a composite of many organs, each with its own functions and its own needs for food, oxygen, and chemical elements. If its many parts operated independently, only chaos would result. But the central control devices so efficiently and so perfectly interrelate that each organ serves the interest of the others and everything works harmoniously.

So precise and reliable are today's mechanical devices that we sometimes remark that they are "almost human."

Such an idea harks back to the time when anything requiring reliability called for a human presence—when a man stood at each railroad crossing to prevent horseless carriages from running into the path of an oncoming train, when the temperature of a building was controlled by the furnaceman, who shoveled more coal into the furnace when the building became cool, and when human hands did the work now done by machines.

Today we entrust the details of routine operation to

Temperature Control

mechanical equipment. We depend on electronic control devices to make adjustments that are more exacting than a human operator who trusts only his five senses can possibly perform. We rely upon computers to work complicated mathematical problems.

But even though we live in an automated world, we must not forget that the human brain still controls. The human brain devised the machinery that now does our work. The human brain worked out the intricate circuits by which electronic gadgetry now does much of our thinking. And we ultimately fall back on the human brain to repair and improve the devices which surround us.

The smooth operation of the body's automatic controls so impresses us that we sometimes speak of the "wisdom of the human body." But really, neither the body nor its components possess wisdom.

Just as the modern computer gives evidence of the "almost human" characteristics of the engineers who designed it, so the human body provides innumerable evidences of the wisdom of its Creator. Although we admire the wonderful mechanisms inherent within the body, we owe a greater debt of admiration to the *Creator* than to the *thing created*. Thus we declare with the psalmist, "I will praise thee; for I am fearfully and wonderfully made: marvellous are thy works."

YOUR BODY'S DEFENSE SYSTEM

Along many modern highways strong barriers protect against cars that go out of control. Some of these surround the trunks of trees growing beside the road. Some divide main highways from parallel roads carrying local traffic. Some separate traffic moving in opposite directions.

The human body has barriers to protect it from the onslaught of the ever-present germs in our environment. Although not made of wood or steel or chain-link fence, they effectively stand guard to preserve the individual's health.

In order to develop an illustration that will help you grasp the importance of these unseen defenses against germs, let me amplify the reference just made to barriers along the highway.

Let us picture the extensive property of a wealthy landowner as it lies in a fertile valley, bounded on one side by a main highway. The man fanatically tries to prevent cars and people from coming on his property. So he has had a heavy board fence built all along his boundary on the highway side. Reinforced by cables to prevent ordinary vehicles from breaking through, it also has fine-mesh wire at the bottom to keep small objects from rolling underneath. Along the top, strands of barbed wire discourage intruders who might try to sneak over.

Inside the barricade this unusual owner has provided

tow trucks and patrol cars in which mechanics and private police constantly traverse all parts of the property. Should an unruly vehicle from the highway happen to break through the fence, the police promptly take the driver into custody and a tow truck hauls the vehicle away to the private wrecking yard hidden behind a clump of trees.

As though this were not enough, this unusual man has installed a giant kennel where police dogs are trained to work with the policemen. These intelligent dogs learn ways to develop their memories, and their trainers boast that once a dog has encountered an intruder, he will always remember the scent of this particular person and will viciously deal with him should he offend a second time.

This strange landowner also employs a crew of fence builders with all necessary equipment at hand. These have instructions, in case the fence is seriously broken, to build a new, temporary fence around the damaged area in order to protect the property from continued intrusion until the trespassing vehicles have been destroyed and the main fence repaired.

In this fabulous system of protection, the main fence along the highway represents the body's skin, with its ability to resist the penetration of germs into the underlying tissues. The wreckers and private police are the leukocytes (white blood cells) which move from place to place throughout the body and engulf germs and viruses, even destroying some of them. The private wrecking yard in which invading vehicles are dismembered typifies the body's lymphoid tissue, including the lymph nodes and spleen, which detain and destroy disease-producing germs. The fence-building crew represents the automatic tissue response that "walls off" an infected area by building a layer of dense tissue which is relatively impervious to the passing of germs and thus helps to prevent the spread of local infection. The police dogs with a memory suggest the mysterious custom-built antibodies which combat unfriendly organisms.

In some cases antibodies do not destroy the germs, but identify them in a way that makes them susceptible to destruction. Furthermore, when a new kind of germ, or virus, such as the virus that causes chicken pox, enters the body and causes an illness, so many antibodies form that some remain after the invaders die out. They remember this particular kind of germ or virus and stand ready to fight it to the death without giving it a chance to cause the same illness a second time.

The unbroken skin forms a very effective barrier against the invasion of germs into the deeper tissues. But once it is broken, germs have access to the exposed tissues and the wound becomes infected.

The body's moist membranes, such as those that enclose the eyes, do not provide as effective a barrier against germs as does the skin. In the case of the membranes of the eyes, an added protective device under favorable conditions keeps these membranes from becoming infected. A continuous flow of clear fluid across the surface of the eyes tends to flush away germ-laden dust particles, rinsing them through the tear duct into the nasal cavities. This fluid, produced by the tear glands, also contains a mild germicidal chemical, the enzyme lysozyme.

Even the air a person breathes is filtered as it enters the body. First, it must pass by the hairs at the entrance to the nostrils, which entrap the larger dust particles that often as not carry germs. Mucus covers the membranes lining the nose and the other air passages. Built-in glands produce this mucus to which the smaller dust particles adhere as the inhaled air moves toward the lungs. These membranes also aid

Your Body's Defense System

in the defense against germs. On their surface millions of tiny fingerlike structures wave gently like wheat in a breeze. These cilia, however, move faster in one direction than in the other, the quicker wave being in the direction of the pharynx. The cilia, as they all wave in unison, carry the film of mucus, which covers the membrane, toward the throat, where it can be expectorated. As the mucus flows along, the entrapped dust particles, including the germs which cover them, move along, too.

Germs frequently contaminate the food which enters the mouth. But there is provision for destroying these before they have opportunity to enter the body's tissues. First, the saliva contains lysozyme, as do the tears. Then the tonsils and related lymphoid tissues located in the pharynx, just back of the mouth, lie in such a position that all food must come in contact with them as it is swallowed. These lymphoid structures produce custom-built antibodies to combat the particular germs present.

Another provision for destroying the germs as they enter with the food depends upon the destructive action of the stomach's strongly acid gastric juice, which sterilizes the stomach contents.

In spite of the means of protection just described, some germs find their way into the body's tissues. Here they move about in the blood, in the lymph stream, and in the tissue spaces, where they prepare to rapidly multiply.

But specialized "police" cells destroy the germs which enter the tissue spaces. Some police cells circulate with the flowing blood. Others anchored here and there in organs like the lymph nodes and the liver come in contact with germs that move through these organs. Some police cells are capable of a simple type of locomotion and can be observed through the microscope seeking out invading germs.

When a pimple or a boil develops in the skin of one's arm, for example, police cells come into the involved area from different parts of the body and form a veritable cordon about the infected area. Squeezing a pimple could rupture this connective-tissue barrier, thus spreading the infection to adjacent areas.

Antibodies are complex chemical substances, produced within certain cells of the body, which make germs powerless to carry on their usual disease-producing activities or else neutralize the poisons produced by the germs.

Some people have a greater capacity to resist disease than others. Some people have tissues which more actively produce antibodies than others. Further, a person's ability to resist disease may vary from day to day. Other factors being equal, a weary person cannot combat an infection as readily as when he is in the pink of condition. A person is more susceptible to the common cold, for instance, when he has lost sleep or is otherwise below par.

In combating infections that commonly involve the skin, such as produce pimples and boils, a person must depend for the most part on his tissues' ability to produce antibodies right at the time of the infection. Upon finishing their work, such antibodies seem to disappear so that the person remains almost as susceptible to a second invasion of the same kind of germ as he was originally.

But a person can succumb to some diseases only once. When this kind of illness strikes for the first time, some of the antibodies produced in response to the infection either remain in the body's tissues or else the tissues' capacity to produce this type of antibody on short notice improves. At any rate, if the germs of this particular disease should strike again, the body resists so promptly and effectively that the second attack does not develop. We speak here of such

Your Body's Defense System

diseases as smallpox and measles, in which one attack confers permanent immunity.

After becoming aware of the body's capacity to produce antibodies to combat disease, medical scientists contrived ways to stimulate their production without a person's having to suffer even once from the disease. Thus vaccination has saved millions of lives. For persons who take advantage of this provision, smallpox no longer terrorizes. Even diphtheria and polio are now defeated foes—except for persons who neglect the protection which vaccination can give.

In principle, vaccination consists of introducing into a person's tissues a harmless phase of the germ for which protection is desired, or dead germs which still retain their capacity to stimulate the production of antibodies, or the toxins produced by such germs. Stimulated thus, the body produces an adequate supply of antibodies to combat the particular disease should its germs ever invade the body.

We can be very grateful that the body contains several mechanisms which maintain our resistance to disease. But we must not rest easy with the thought that with the body's built-in systems of defense we stand completely secure. Marvelous as these protective provisions are, their degree of efficiency depends upon one's state of health. We each have a part to play in maintaining our resistance to disease. Each person must make sure that he follows a way of life that promotes physical fitness and thus makes his body's defenses most effective.

REPAIR WORK

I am thinking of a terrible spectacle of physical tragedy. He carries his right arm in a sling because he once broke the bone and the fractured ends have never grown back together.

Four Band-Aids patch his face. On his forehead a dark-colored gash extends back into his scalp, and surgeon's stitches hold the edges together.

Almost every finger carries a bandage or has some sort of open wound. Each cut and sore has a history which he willingly recites. He clearly remembers each incident that caused him pain and inconvenience, but he talks slowly, for inside his mouth a painful ulcer which he has had for many months impedes his speech.

He is in the hospital just now having the doctor restitch a surgical incision. He explains that once he had appendicitis and the appendix had to be removed. The incision never healed, and so each time the sutures wear out he returns to the hospital to have new ones put in. "But I am worried," he admits, "because the tissues at the margin of the incision are just about frayed, and it's hard for the surgeon to find a place to put the new stitches."

"A preposterous story," you say.

And so it is, for a person whose tissues lack the ability to heal themselves would soon bleed to death or die of infection. But before we dismiss the illustration, let us ponder

Repair Work

it long enough to admit that this fictitious man represents you and me except for the miracle of healing.

Again at this point the hypothetical case breaks down, for if human tissues could not heal themselves, all mankind would suffer. Doctors and nurses could not stitch the lacerations or splint the broken bones of their fellow sufferers.

As for the accidents and injuries which the man of our imaginary story has suffered, he has had quite a normal experience. Thousands would have just as many evidences of life's minor tragedies did not broken bones knit, the edges of lacerations fuse by the local production of connective tissue, the skin margins of torn or denuded areas generate new cells, muscle tissue splice by the elaboration of a strong and pliable cementing substance, and even severed nerves become functional again as lengthening nerve fibers find their way across an area of tissue repair and use the investments of the old fibers as conduits to guide them to the right locations in the tissues.

A surgeon readily admits that his skill can never rehabilitate one who is injured. His best efforts would accomplish nothing apart from the healing processes that begin where the surgeon leaves off. He sets the stage; the inherent capacity of tissue to accomplish repair produces the final results.

Only plants and animals—those living things which God created—exhibit the marvel of self-repair. Only in these can the blemishes imposed by accident and disease automatically disappear by processes of restoration which keep the tissues in good condition.

With man's inventions, repair proceeds from the outside. A high-quality self-winding watch, one of the most reliable instruments in common use, sooner or later begins to lose or gain time, or it stops running. Then what do we do? No one simplemindedly thinks that if he lays the ailing watch on the shelf for a few days, it will repair itself and begin to keep good time again.

Some manufacturers describe their automobiles as "self-serviced." By this they mean that such a car will run thousands of miles without the care of a mechanic. It has a built-in lubrication system; the brakes adjust themselves; many of the bearings are "sealed" and expected to function well without attention throughout the life of the car.

But no manufacturer has ever advertised that his car will repair itself. What car can smooth out its own crumpled fender? Where is the car that will automatically cover its scratches with new paint of the right shade? What car mends its own upholstery? What car can weld a break in its own frame?

Yet consider how the body repairs a broken bone. Even though bone contains combinations of calcium and magnesium, which make it very hard and strong, living cells occur throughout the bone substance, and these have the same needs as cells located in soft tissues: needs for oxygen, for nutrient material, and for the removal of waste products. So an extensive network of blood vessels penetrates the periosteum (the fibrous surface layer of a bone), the marrow of the bone's interior, and the minute tunnels which run lengthwise in the bone shaft.

The same force that causes a bone to break causes the fragments to separate from each other; and even though they may return to almost normal positions, the network of blood vessels within and around the bone is disrupted at the time of fracture, and blood seeps into the tissues in the vicinity of the break, which triggers the mechanism for repair.

Bone-forming cells which ordinarily lie dormant in the marrow and in the periosteum now become active and in-

Repair Work

vade the tissue around the newly broken ends of the bone fragments. Here they produce spicules of new bone and a layer of cartilage just as their ancestor cells did during childhood when this bone developed and grew.

The newly formed tissue at the site of a fracture is called a callus. X ray shows it to fit over and between the ends of the bone fragments. Containing some cartilage, a callus can flex slightly. But gradually the amount of bone increases and the amount of cartilage decreases, and the callus, serving as nature's internal splint, becomes more and more firm.

When the ends of the bone fragments fit well with respect to each other, as when the doctor has "set" the broken bone and checked it by X ray, the amount of the callus is relatively small. When an "offset" mars the position of the fragments, a larger callus must envelop the portions of the bone needing repair.

As mentioned, a callus is flexible when first formed; hence, the doctor in charge makes provision for keeping the broken bone stationary until the callus becomes firm and strong enough to hold the fragments in good position. So he uses an external splint or a cast to hold things in place for a while.

Nature's process of bone repair does not stop with the formation of a strong callus. A callus is bulky and appears something like the bulge produced when a boy fastens two sticks together by winding string around their junction point. So, once a callus grows strong enough to permit the normal use of the injured bone, a new kind of tissue activity begins among the cells contained within the callus. The excess bone material is destroyed and what remains alters so that it conforms more, both in general appearance and in internal architecture, to the structure of the bone before the break. Eventually, even the X ray shows very little evidence that the bone was once broken at this site.

Next we consider the method of automatic repair which the body uses following a burn.

Normally the cells in the deeper part of the epidermis (the cellular, surface layer of the skin) actively produce new cells. In this layer of cell activity, one cell splits to make two identical cells, and as the number of cells increases, many cells push toward the surface. As they approach the surface, they lose contact with their source of nutrition, become horny in texture, and break away from the skin surface as newer cells replace them.

A mild, so-called first-degree burn does not destroy the cells of the epidermis; it only injures them. The skin reddens because of the irritating effect of the heat, and consequently the blood capillaries in the dermis (the second, deeper layer of the skin) enlarge. The pain of a first-degree burn is due to irritation of nerve endings in this deeper skin layer.

A second-degree burn destroys the cells in the epidermis, which, even though now dead, may remain intact for a while, but they usually detach from the underlying dermis, permitting tissue fluid to infiltrate between the two layers, forming a blister. In this case, healing takes place by increased activity of the cells at the margins of the area of cell destruction. These cells subdivide actively to produce new cells, some of which push toward the surface of the skin in usual fashion while others move toward the center of the blistered area—or onto the raw area if the blister has already broken.

A small area of second-degree burn usually heals quickly and leaves no permanent scar. A larger area may require a longer time for the newly formed cells to reach far enough over the injured area to join those that have been approaching from the opposite margin. In such a case, small skin grafts here and there over the raw area speed up the healing process. Each of these, as it becomes established, serves as a focus from which new epithelial cells develop to join those of nearby areas.

In a third-degree burn both layers of the skin (epidermis and dermis) are at least partially destroyed. These deeper burns heal slowly, first, because the intense heat has devitalized the cells at the margins of the injury and, second, because the wound is susceptible to infection. With the dermis injured as well, a certain amount of healing here must form an appropriate tissue foundation on which the cells of the epidermis can rest. The dermis, composed mostly of connective tissue, heals primarily by the reinforcement of this connective tissue in a manner that produces a scar. Once the dermis layer heals, then the active epidermal cells at the wound margin can produce new cells to cover the defect. Even so, the scar will show through the new layer of epidermis, leaving a permanent disfigurement.

We have described only two of the body's many provisions for repairing its tissues. We could write an entire book on the subject, detailing the manner in which broken blood vessels mend, the method of repair in tendons, the way the broken ends of muscles unite, the process by which severed nerves are restored to function, and many others.

We should also mention an even more important form of healing—it affects a person's eternal welfare. We refer to the kind of healing the psalmist mentioned when he wrote, "Lord, be merciful unto me: heal my soul; for I have sinned against thee." Psalm 41:4. The same God who created our bodies and made provision for the repair of injuries to their tissues has abundantly provided, through the plan of redemption, to erase the effects of sin and heal our souls.

His desire that everyone avail himself of this spiritual

healing prompted Jesus' comment when, during a Sabbath sermon in His home church at Nazareth, He stated the reason He had come to the earth: "To heal the brokenhearted, to preach deliverance to the captives." Luke 4:18.

SPEECH: A MARVEL OF COORDINATION

One afternoon I strolled through an aviary filled with many varieties of birds. After looking at many of them, I came to a cage with a single bird. As I looked this way and that to find a marker giving its name, the bird came over to me and startled me by saying, "I'm a myna bird."

Even though I was taken aback by the bird's clever trick, it did not convince me of its superior intelligence. Instead, I gave credit to some playful caretaker for having drilled the bird on what and when to speak.

Birds that speak are not necessarily more intelligent than those that are mute. Neither are birds—even those that speak a few words—better able to "use their heads" than forms of animal life such as dogs and horses which often manifest enough intelligence to do the right thing at the right time.

Why, then, in our appraisal of our fellowmen, do we so commonly use what a person says as an index of his intelligence? Does his ability to speak serve as a medium for conveying his thoughts? We judge his intelligence not so much by how much he speaks as by the quality of the thoughts his speech transmits.

Speech has two components: one is mechanical and has to do with the production and control of sounds; the other is intellectual and involves the use of sounds as symbols to convey clearly to others one's innermost thoughts.

The first requirement for the production of speech is power—power to activate the vocal cords. At first mention we are tempted to say that the column of air leaving the lungs, by way of the windpipe, rushes through the larynx (voice box) and past the vocal cords, producing speech. Yes, the moving air causes the vocal cords to vibrate and produce sound. But what causes the air to move? Fundamentally the power required to move air out of the lungs and thus to produce speech derives from the action of muscles.

Two sets of muscles produce respiration, one bringing air into the lungs and the other forcing it to leave. The diaphragm and the muscles between the ribs, when they contract, increase the capacity of the chest and thus cause air to rush into the lungs. Once the lungs fill with air, a mere relaxation of the inspiratory muscles allows a considerable amount of air to leisurely escape.

Forced expiration, in which significant pressure pushes the column of escaping air, is caused by the contraction of the abdominal muscles. Prove this to your own satisfaction, if you like, by pressing the tips of all your fingers firmly against your abdomen while you take a full breath and then expel the air as quickly and forcefully as possible. You will notice that the abdominal muscles firm up as you force the air out.

But the muscles of respiration do not contract on their own. In this they differ from the muscle of the heart, in which contraction follows contraction whether or not the nerve impulses stimulate the heart muscle. The muscles of respiration contract only when they receive nerve signals to do so—like the muscles of the arm or of the leg.

At this point you will probably raise a question, for you have observed that the muscles involved in breathing work

Speech: A Marvel of Coordination

almost automatically, day and night, with those of inspiration alternating their contractions with those of expiration. The muscles of the arm and leg move on demand as you bring them into service in carrying out your wishes. This is the reason they are called "voluntary" muscles.

But let me explain that the muscles of respiration and the muscles in the arm and leg resemble each other more than it may seem. They are all properly classed as voluntary muscles. Your will controls the actions of the muscles of respiration, up to a point. You can hold your breath or you can speed up the sequence of taking one breath after another.

Actually, the muscles of your arms and legs do not require conscious attention each time they contract, either. When you decide to take a walk, you do not have to think about muscles each time you take a step. You simply determine which direction you will go and how fast and then turn over the job of controlling the muscles to certain coordination centers in your brain.

A similar method activates the muscles of respiration. A "respiratory center" in the medulla portion of the brain sends out intermittent nerve impulses. This respiratory center is easily influenced either by the conscious centers of the brain or by the body's need for more or for less oxygen.

Also, the momentary needs for air to produce the sounds involved in speech affect the respiratory center. When a person speaks, coordination centers in the brain induce an interruption of the usual breathing cycle. Instead of one breath following another at a steady pace, a gulp of air is now taken quickly into the lungs and then expelled gradually under pressure as may be needed by the larynx. The singer sustaining his voice until the end of the musical phrase affects his breathing even more markedly than when speaking.

Thus far we have discussed the muscles of respiration. In addition to these, the small muscles within the larynx affect speech through their influence on the position and tautness of the vocal cords.

In ordinary quiet breathing the vocal cords relax and lie folded against the walls of the larynx in order not to obstruct the flow of air on its way to and from the lungs. When a person holds his breath, he does so by bringing the vocal cords together, one from each side, so that they meet each other and prevent the escape of air from the lungs. Pressure in the column of escaping air is produced by bringing the vocal cords close enough to each other to reduce the volume of air that escapes while the abdominal muscles contract.

The vibration of the vocal cords as the air passing between them agitates them produces the sounds required for speech. The more taut the vocal cords as they vibrate, the higher is the pitch; the greater the pressure of the column of escaping air, the greater the volume.

Just as with the muscles of respiration, so nerve impulses activate the muscles of the larynx. And here again centers of coordination in the brain manage the impulses that go to the muscles of the larynx and bring about delicate changes as needed. Just as in the control of walking or of other skilled movements, a person is not consciously aware of the action of the individual muscles. He only senses his desire to make certain sounds. The brain's centers of coordination do the rest.

The vibration of the vocal cords produces relatively simple sounds, comparable to those produced by drawing a bow across the strings of a violin. Therefore, word pronunciation requires a modification of these simple sounds, and the muscles of the palate, the tongue, and the lips accomplish this. These structures cut off, release, or impinge on

Speech: A Marvel of Coordination

the column of escaping air to produce the consonant sounds and control the factors of resonance to perfect the vowel sounds.

We understand now that speech involves three groups of muscles: the muscles of respiration; the muscles within the larynx; and those of the palate, tongue, and lips. Each of these muscle groups performs other functions besides those relating to speech. The muscles of respiration are concerned twenty-four hours a day with providing the lungs with a constantly changing supply of air. The muscles of the larynx, in addition to adapting the vocal cords to produce sounds, have to do with regulating the pressure of the air within the lower air passages as when involved in coughing or in holding the breath when lifting or straining. The muscles of the palate, tongue, and lips are active in masticating and swallowing food and drink. A coordinating center within the brain controls each of these groups of muscles so that they can carry on their ordinary tasks smoothly.

When it comes to speech, however, a super coordinating center must harmonize the three muscle groups. This "speech center" is located at the highest level of integration of the brain—in the cortex of the cerebrum. From this vantage point it dominates the other coordinating centers already mentioned so that speech takes priority, if need be, over other muscular activity.

The cerebrum of the brain consists of two hemispheres, a right and a left. For many functions the right hemisphere controls the left side of the body and the left hemisphere, the right side of the body. Not so for the control of speech, however, for it is a unified function intended to play an important role in the intellectual processes. There cannot be two speech centers, one in the right hemisphere and one in the left. So the one speech center resides in the "dominant" hemisphere, which, in a right-handed person, is the left hemisphere.

The speech center serves as more than a mere control board. It receives nerve connections from the nearby parts of the brain in which ideas develop and in which consciousness exists, and when the time comes to transmit an idea to someone else, the speech center "programs" the necessary actions of muscles and their sequence. Nerve impulses carry this pattern to the centers in the lower parts of the brain which initiate the muscle actions.

Without a properly functioning speech center in the brain, the sounds produced by the organs of speech could be compared to the sounds we hear when the musicians in an orchestra tune their instruments. We hear all varieties of sound but no music. The influence of the speech center parallels the influence of the orchestra conductor when he takes the baton. It brings order out of chaos so that the sounds coming out of the mouth follow a meaningful pattern and thus serve as symbols to convey thoughts and ideas.

In the present discussion we have used speech as an outstanding example of those important abilities possessed by human beings which various combinations of nerve cells, muscles, and other tissues make possible. We marvel at the Creator's handiwork. But even more remarkable than the complicated structure of the organs of speech is the provision for their coordinated control so that in producing the spoken word they make it possible for a person to transmit his thoughts to someone else.

Speech is only one of many functions of the human organism which display the Creator's wisdom. The intricacies of the provision for hearing and the ability to attach meaning to the spoken word once it is heard are just as profound.

Speech: A Marvel of Coordination

Writing as well as speech conveys thoughts from one person to another. Here the brain's intellectual centers influence the actions of muscles in the hand and arm rather than in the organs of speech. And when a person reads what has been written, another marvel becomes effective—the ability to see the written symbols and comprehend the meanings for which they serve as a vehicle.

One's words reflect his thoughts and attitudes and thus provide an insight into character. Notice how the psalmist made application of this relationship in an appeal for personal nobility: "Let the words of my mouth, and the meditation of my heart, be acceptable in thy sight, O Lord, my strength, and my redeemer." Psalm 19:14. The Master, when on earth, also recognized that words reflect character and reminded us of our personal responsibility for the words we utter: "By thy words thou shalt be justified, and by thy words thou shalt be condemned." Matthew 12:37.

HOW YOUR BRAIN KNOWS

Thought and consciousness occur in the brain. Memory, imagination, and the exercise of judgment all depend upon complicated activity involving nerve impulses within the brain.

Although like an electronic computer, the brain is much smaller, of course, much more adaptable, and much more efficient than the commercial computers now in use.

A computer has to be provided with data. When a scientist uses a computer to solve some problems, he must "program" the data so that the computer has information on which to base its activities.

Similarly, the human brain must receive information in order to operate. And this information reaches the brain in the form of nerve impulses carried by bundles of nerve fibers which pass through their respective openings in the skull to reach the brain.

The brain's awareness of its surroundings depends upon the nerve impulses that come from the sense organs which God designed to register a particular kind of information. The eyes deal with light, color, and form; the ears, with sound; the taste buds, with the soluble chemical substance in one's food and drink; the organ of smell, with the volatile substance contained in the air or the food; the organs of equilibrium, with changes in the position of the head; and the organs of body sensation, with the physical changes that take place in the skin and in other tissues of the body. In every case a sense organ transforms some kind of energy into nerve impulses in a manner which the brain "programs" appropriately.

The eye resembles a camera. It has a lens just behind the circular colored area, or iris, which operates like the diaphragm in a camera, admitting more light or less, depending on circumstances. The opening at the center of the iris (the pupil) appears black because a dark pigment coats the inside of the eye; it prevents reflection just as the black paint inside a camera keeps everything dark.

The very front portion of the eye, the cornea, is transparent, and you can see it best by observing the eye from one side. It is shaped like the crystal of a watch, and a clear fluid behind the cornea fills the space between it and the lens. The iris "hangs" in this fluid-filled space.

In a camera, the image in front focuses on the film at the back of the camera. In a similar manner, the eye focuses the visual image on the retina, at the back of the eye. Two kinds of light-sensitive nerve cells—rods and cones—occur in the retina. Each eye contains about seven million cone cells and about 125 million rod cells. The cone cells register detail and color. The rod cells aid vision in reduced light and register movement within the visual field.

Through nerve fibers each rod and cone cell contacts the area of the brain which perceives vision. As a visual image focuses on the retina, some of the cells register dark and some light, depending upon the particular pattern of the image. Some of the cone cells register red, some green, and some blue—the primary colors to which these cells are adapted. Combinations of these provide all other shades and color combinations.

Each eye "sees" the same object as the other eye, but from a slightly different viewpoint. The impulses sent back to the brain from the right eye vary slightly from those sent by the left eye. The brain blends the two images, and this accounts for the ability to detect depth and perspective. There is more advantage, then, than merely having a "spare," to the arrangement by which we have two eyes rather than one.

The eye is so important to the individual's well-being and personal pleasure that the Creator adequately protected it by setting it within a bony chamber, the orbit. At the front the upper and lower eyelids come together, reflexly, whenever danger threatens and guard it. More than this, the tear glands, located beneath the skin in the outer portion of each upper eyelid, produce a small amount of fluid even when a person is not crying. As the eyelids close during each blink, they spread this fluid over the surface of the eye and so keep the exposed tissues moist. This fluid even contains a germicidal substance.

The eye marvelously adapts to changing conditions. When light abounds, delicate circular muscle fibers in the iris contract, causing the pupil to become smaller so that less light enters. At the same time, the light-sensitive substance contained in the retina chemically changes, thus preventing what the photographer would call an overexposure. In twilight conditions, the pupil becomes larger and the retina more sensitive to light so that a person can see remarkably well even in semidarkness.

The ability to see at close range serves as another example of the eye's capacity to adapt. A photographer has to adjust the lens of his camera in order to bring a close object into sharp focus. In the eye, small strands of muscle change the tension on the lens, causing it to become thicker and thus capable of focusing light from a close object. Strange as it seems, completely transparent living cells and tissue fibers compose the lens. Remarkably, this lens produces a visual image well-nigh perfect in its detail.

The ears are adapted to deal with sound waves rather than light waves, and we can compare the inner ear to a musical instrument just as we likened the eye to a camera. But in comparing it to a musical instrument, we must recognize that it is *receiving* sound rather than *producing* it.

If you are familiar with the piano, you know how it resonates when you depress the pedal. Then, even though you do not touch the keyboard, the strings of the piano will vibrate in unison with any musical note such as that produced by the human voice in singing. A high-pitched note sung into the piano will cause the treble strings to respond and "sing back" a note of the same pitch. To a low note, the bass strings respond similarly.

The inner ear does not have long strings as does a piano, but rather a series of strong connective tissue fibers arranged in order, from those which respond to high-pitched sounds to those which respond to very low-pitched notes. In order to conserve space, the inner ear's "soundboard" spirals like a snail shell.

When the internal ear receives vibrations of a given frequency, the fluid which this organ contains quivers. This causes a specific part of the soundboard to vibrate in unison with the sound vibrations. In so doing, receptor cells located here relay to the brain nerve impulses which it interprets as a sound of a particular pitch.

But most sounds received by a person's ear are complex. The sound of the human voice has more components, even, than a musical chord of several notes. This means that listening to music or to the voice of a friend involves several parts of the organ of hearing. So the portion of the brain devoted to the interpretation of sounds receives a whole volley of nerve impulses coming from various parts of the organ of hearing and representing sounds of various pitches.

Out of this the brain reconstructs the information brought to it in the form of nerve impulses. If a spoken word is heard, the brain attaches significance to the combination of sounds so that the person understands the meaning of the word he has heard. In the case of music, the combination of various pitches of sound is pleasing. In ordinary "noise" the brain may respond to the combination of sounds with disinterest or annoyance.

Fortunately we have two ears rather than one. Many persons get along very well after they have lost the hearing of one ear. But just as two eyes enable a person to see depth and to judge distance, so the two ears make it possible for a person to judge the direction from which a sound comes.

When a given sound wave arrives at the right and left ears at exactly the same time, the brain interprets this to mean that the source of the sound lies directly ahead (or directly behind). But if it arrives at one ear slightly in advance of its arrival at the other, this indicates that the source of the sound is to the right or to the left, depending upon which ear receives the sound wave first. Often when looking into the sky or across a landscape, we enlist our ears to help us locate an airplane or the position of a bird. Then, having determined its approximate location by hearing, we make a further search by the use of our eyes.

Sounds may be loud or faint, high-pitched or low-pitched. In attempting to discern a faint sound, a person automatically turns his head so that the sound enters his ear directly, and he may cup one hand behind the external ear to increase its sound-gathering advantage.

How Your Brain Knows

At the other extreme of the scale, a person may place his hands over his ears to protect them from the overstimulation of a very loud sound. But a built-in safety mechanism also cares for this.

The sound waves which enter the external ear cause the eardrum, located at the depth of the auditory canal, to vibrate. On the other side of the eardrum three tiny bones (ossicles) carry the vibrations of the eardrum across the middle-ear cavity to the inner ear, where the fluid which surrounds the receptor cells starts vibrating. In this middle ear, where the bony ossicles are located, two small muscles, when they act, dampen the movements of the ossicles and thus prevent damage to the ear's delicate structures. These muscles go into action reflexly without the individual's knowledge or supervision. They exemplify the safety devices that the Creator has placed in the body.

Although the sensations of smell and taste do not provide the brain with top priority information—as in the ability to read or understand spoken language—they do send messages which add to the pleasures of living and also, in some cases, provide warnings of danger. The organ for smell, located high in the nasal cavities, samples the air breathed or registers the nature of the volatile components of food eaten. The taste buds function in a similar manner by registering the kind of food or drink taken into the mouth, and by sending to the brain a report on whether the substance tastes sweet, salty, sour, or bitter.

Scattered throughout the skin and the membranes that line the body's tubes and cavities, as well as in the joints and muscles, many small receptor organs which are served by nerve fibers respond to touch, heat, cold, pain, and pressure. No one receptor can register all of the sensations just listed; a variety of sense organs receive their own kinds of sensation.

As a result the brain senses whether its tissues are in favorable circumstances.

Although each of the organs of special sense is marvelous in itself, the way in which the various sense organs relate to the total function of the body and to each other to make life enjoyable as well as worthwhile is even more remarkable.

No better statement exists on the importance of the several parts of the body, as they contribute to a human being's welfare, than that of the Apostle Paul: "If the ear should say, 'Because I am not an eye, I do not belong to the body,' that would not make it any less a part of the body. If the whole body were an eye, where would be the hearing? If the whole body were an ear, where would be the sense of smell? But as it is, God arranged the organs in the body, each one of them, as he chose. If all were a single organ, where would the body be? As it is, there are many parts, yet one body." 1 Corinthians 12:16-20, R.S.V.

YOUR BRAIN KEEPS RECORDS

We had such a good time at our house the other evening just showing pictures on the screen. Everybody enjoyed it, even though the pictures were not unusual or new. In fact, that's why they interested us so—they were not new. I took them twenty years ago when the children had just entered their teens.

One showed Ed with his long arms and legs sticking far out of sleeves and pants legs. In another Patricia sported a funny hairdo.

"Mother," one of the granddaughters shouted, "why in the world did you fix your hair that way? It looks terrible."

"We thought it was the greatest in those days," Patricia hastened to explain.

"Where did you get those pictures, anyway?" my wife queried.

"Oh, I had them in the cupboard," I explained. "They have been there all the time, but it's been years since we have looked at them."

Just as my cupboard is full of pictures—some taken last month and some taken years ago—so my brain stores many memories, some relating to things that happened only last month and some to what occurred years ago.

You would think the "memory department" of a human brain would become cluttered. Every day it records many events. Thousands of new memories are established every year. By the time a person has lived a few decades it would seem that his memories would be in terrible disorder.

Disarray marks the cupboard in which I keep my pictures. Here is a group of pictures that I pulled out of their box to show some visiting friends. I had taken them when we visited their country home. After we looked at them, I put them aside, and now I have mislaid the box in which they belong. Here is another group with no writing on them. They show trees, water, and some buildings in the distance. I can't tell where I took them, and nothing identifies the occasion. There are no clues as to whether they picture California or Oregon. What a mess!

Not so with memories, however, for the brain has a marvelous capacity to keep its memories well organized and available on a moment's notice. Granted, many memories of routine things fade out, perhaps to make room for more important items, but for the significant or interesting memories the brain possesses an intricate cross-reference system. One memory links with another by "associations" that serve to unravel the whole story, item by item. Here's an illustration of how it works:

Perhaps you wish to recall the name of a friend you met on vacation last summer. "I can see his face in my mind's eye," you say to yourself, "but his name escapes me. I remember that he spoke about the trip he had made up to that time. He said he had stopped to see some relatives whom he had not seen before. He told of the hard time he had finding them and of how he looked in the phone book and said that it listed many people by the same name. Their name was the same as his. I recall that I wrote down his name and address so that I could correspond later. His name is the same as another friend of mine, and I remember comparing the two persons—this new friend is taller and more talkative. Now it comes to me; I compared him to Jim Roberts. Roberts—that's the name, Kenneth Roberts. 'Just call me Ken.' I can still hear him say it."

Another remarkable fact about the human brain and its ability to keep a record of past events is that it utilizes such a small amount of space. When I compare human memory with my library of colored pictures, I just cannot understand how the brain can keep on remembering new things without needing more space in which to store the increasing amount of memory material.

The cupboard in which I store my slides became full several years ago. Every year I have another batch of slides to keep with no place to store them. So I have left them in their original boxes and have piled these on the deck just outside the cupboard. Now when I hunt for some particular picture taken several years ago, I have to lift off all the boxes of recent slides before I can open the cupboard door. I surely need a new and larger cupboard.

But the human brain seems to thrive on having more and more material to pack away in its archives. It is marvelous how an organ which weighs only three pounds can go on year after year storing new memories while still keeping the old ones available. And the marvel becomes even greater as we remind ourselves that the task of remembering is only one of the brain's many activities.

The other day I watched the gardener through my office window as he raked up some leaves. No one else was around, and I didn't reveal my presence. Gradually, as he kept on raking, a smile crept over his face. At first he puckered his lips rather self-consciously and tried to look sober again. But soon the smile broadened, and it seemed to me he almost laughed out loud.

He wasn't talking to anyone. He wasn't watching anything amusing—just following the strokes of his rake. What made him smile?

Being all alone, he drew on his memories for material to occupy his thoughts. As he rehearsed one memory after another, a funny one came along. As he relived this particular experience of the past, he reacted to the memory just as he had reacted in the original situation—with amusement.

Memories often provide entertainment. Some amuse. Some sober, and some actually sadden. Whatever the original experience, the memory of it re-creates the same emotional response. So with my colored pictures. When flashed on the screen, they enable those of us who were there when I took the picture to relive the whole experience.

But memories do more than entertain. Entertainment is only a small part of the purpose for which the Creator endowed us with the capacity to remember. Memories give us a flowing contact with the past so that we can pick up today where we left off yesterday. We don't have to make our mistakes over again. By using our memory we can say, "My experience teaches me that I can best solve this problem by reacting differently from the way I did the first time."

Furthermore, memory provides a growing knowledge of how to do things efficiently. The first time you made out your income tax report, you found it difficult. The next time it was easier because you remembered methods laboriously worked out the first time. And it becomes easier each time because, by relying on your memory, you pick up where you left off last time.

I find that my interest in taking pictures runs in cycles. Sometimes I am very camera conscious and try to take pictures of all events that interest me and the family. Then at other times I forget to take the camera or I run out of film and so fail to get "shots" of important happenings. Thus my sketchy set of pictures provides an incomplete record of what has taken place in our lives.

Not so with memory. Rather like a TV news camera, it constantly operates. It records everything that takes place. In fact, the records which the brain keeps are even more complete than those made by a TV camera. Memories include sight, color, and sound, just as are captured by modern recording equipment; but in addition, memories record smells, tastes, attitudes, and emotions. The entire experience through which a person passes makes its impression on the cortex of the brain.

I notice that some of my older color pictures, particularly those made during World War II with inferior dyes, have faded a bit. Also, some that I have used over and over again do not appear as attractive on the screen as they used to. Perhaps the projector's strong light has caused them to deteriorate. I have said for a long time that I must sort through my pictures and throw out those no longer satisfactory.

Memories, too, tend to deteriorate, but not from overuse. Quite the opposite. The frequently recalled memory remains most vivid, whereas the memories allowed to lie idle gradually become vague.

In order for a memory to become permanent, it must be recalled from time to time. Thus occasional review is listed as one of the principles of learning. The student reads his assignment today and establishes memories of the facts contained in the lesson. Then, if he is wise, he reviews the assignment briefly after two or three days. This activates the memory patterns in his brain and has the effect of making them more permanent. Still such memories tend to fade if not put to use. So the good student rehearses them again just before examination day.

We marvel at the near-perfect memories which some people possess. Perhaps we listen to a lecture delivered by a historian who recites dates and quotes references without referring to notes, and we say, "What a memory!"

This is no more remarkable, however, than for a housewife to put various amounts of several ingredients together in just the right order and at just the right time as she prepares a delicious recipe for the family's fare and does it all from memory.

The historian remembers what he does because it forms part of his life's enterprises. The housewife remembers the complicated recipe because she has attended to its details on previous occasions, and now her memories of it are completely reliable. In other words, a person can remember what he wants to remember.

We have considered various ways in which God's love, foresight, and wisdom manifest themselves in the human body—the masterpiece of His creation. Now, as we have just explored the remarkable gift of memory, we find in it another evidence that He has built into our bodies those features and capacities which contribute to our physical well-being and our mental satisfactions.

God deals with us as intelligent beings. Although many parts of our bodies perform their functions automatically without direction from our conscious thoughts, the brain can respond independently. Hence, our thoughts are what we make them. Our memories consist of the items that we have cherished the most. By individually exercising our power of choice we determine our way of life. God does not force us to act. He allows us to determine whether we will respond to the way of life that He has so clearly revealed in the Bible. But He does give us instructions that, if we choose to respond to them, will enable us to fulfill His purposes for our lives. He even advises us regarding what we should remember.

"Remember the sabbath day, to keep it holy" (Exodus 20:8), He invites. "Remember now thy Creator in the days of thy youth" (Ecclesiastes 12:1) also refers us to our memories. And in speaking to the members of the seven churches as portrayed in the Book of Revelation, He appeals for repentance in the words, "Remember therefore from whence thou art fallen, and repent." "Remember therefore how thou hast received and heard, and hold fast, and repent." Revelation 2:5; 3:3.

MOVIES OF THE MIND

Your mind has a most fascinating built-in motion-picture screen, on which you can see realistic or fantastic pictures, whichever you choose. By a whim you can put on programs of your past history including action, color, and even the voices of those who participated. These historical plays we call *memory*.

Also, on this magic screen contained within your brain, you can see futuristic scenes of marvels that have not yet happened. This interesting form of mental activity, which we call *imagination,* holds no limit to the splendor or the horror (as you choose) of what may appear on the screen. The only requirement for this fantastic mental drama is that you have a bit of leisure so that thoughts related to your activity of the moment do not jam the picture which your imagination provides.

The program which your imagination concocts blends, combines, or reproduces in new settings the material filed away in the archives of your memory. Some of this "stuff" which contributes to the scenes which flash on your personal viewing screen comes from actual experiences you have had, some from conversations with other people, some from what you have read, and some from what you have seen in pictures.

Imagination and memory use the same "screen" by

which the "mind's eye" sees (and the "mind's ear" hears). Both imagination and memory can be so realistic and the picture on the screen so vivid that you almost feel as if "I am there."

One difference distinguishes between imagination and memory: memory pertains to the past and imagination refers to the future. Memory reproduces or reviews things that have already occurred. Imagination, by contrast, embraces fantasy and depends on make-believe.

Daydreaming is one of the commonest and simplest examples of imagination and typically begins with, "I wish——" or "Suppose——" Ordinary daydreams are usually pleasant, and why not? The imagination produces whatever kind of program you specify.

Daydreams are very personal experiences. No one else except the one doing the daydreaming sees the picture on the mental screen. The censor is one's own conscience, and no other audience passes approval or disapproval.

The content of your daydreams indexes your character. If you want to determine the quality of your character, you do not have to take tests or fill out complicated questionnaires. Simply recall your last few "air castles," and notice whether they were selfish or unselfish, vulgar or pure, cowardly or courageous, pessimistic or optimistic, full of doubts or full of courage.

Indulging in imagination provides such an interesting pastime that one can easily spend too much time watching the screen in his own brain. This overindulgence in make-believe especially tempts teen-agers, who are curious about the world around them and particularly about the way they will relate themselves to future events. This causes them to reach out and try to picture themselves in an adult world, in situations which they have not yet experienced. Boys and girls in their teens like to see on their brain's screen pictures of the future relating to marriage and dealing with their supposed successes in life's undertakings. Such daydreaming is natural and perfectly proper so long as the teen-ager does not carry this passive experience to extreme. In fact, moderate castle-building even helps the teen-ager to establish his preferences and to develop his plans for the future.

Many a person expresses surprise, when he reaches adulthood, to find that the air castles he built while in his teens have now come true. How? Not because he had any prophetic gift by which, when a teen-ager, he could read his future, but because air castles represented his first preferences. As life unfolds, he tends to follow the dictates of his cherished air castles. Again we see that a person's daydreams give a true index of his inner thoughts and desires. No wonder King Solomon said, "As he thinketh in his heart, so is he." Proverbs 23:7.

Unhappy people have a special tendency to overindulge their imaginations. When actual life is not pleasant, the person who feels thwarted easily builds mental pictures of the way he wishes life could be. In this kind of daydreaming, the disappointed person is always the hero. In his daydreams he accomplishes great things—things he cannot accomplish in reality. Such a person derives his pleasures and satisfactions more from his daydreams than from the things he carries forward in real life. Such an unhealthy situation robs the individual of the zest and determination necessary to overcome obstacles and to succeed in spite of hardships, and he begins to reason, "Why should I knock myself out in the effort to succeed when I can find just as much pleasure sitting in an easy chair, enjoying my own thoughts?" Because too much daydreaming causes one to be content with what his imagination produces rather than forcing himself toward success in real life, every person should exercise self-restraint and not allow himself to indulge in daydreaming to excess.

Imagination forms the proving ground on which a person tries out his new ideas. The sculptor always builds his masterpiece in his "mind's eye" before he takes up his hammer and chisel to shape it out of marble. Every composer of music first hears the new melody as a "tune running through his head." Every artist paints his picture first on the screen of his brain and transfers it later to canvas. Every architect "sees" the new edifice before he makes any marks on his drawing board. The interior decorator consults a mental picture of color combinations before he specifies the pigments for the paint or the fabrics for the upholstery. Every good cook, in planning a new dish, consults the taste buds of her imagination before she takes the ingredients off the shelf.

The most satisfactory way to make any decision is to enlist one's imagination in constructing a mock-up of future consequences, and for this the imagination is ideally suited.

When Janet wonders whether to marry Joe or Harry, she should ask her imagination to paint a picture of her circumstances ten years hence should she be the wife of Joe. Drawing on her knowledge of Joe's traits, ambitions, and talents, the screen of her brain will present a fairly good picture indicating her situation should she become Joe's wife.

Then she should change the scene and follow a similar procedure, rolling back the curtain of the future should she marry Harry. With Harry's different personal qualifications, the two pictures will present a contrast. Once she completes them, Janet can sit back, make comparisons, and decide which she likes better.

Each of life's major decisions should be made in this manner. Projecting the mental picture as far as possible into the future should prevent the disappointments that come from rash judgments and from acceding to the human desire to sample pleasures of the moment.

One young husband wrote that his imagination gave him a hard time by causing him to doubt his wife's loyalties. He held an evening job, and his wife, a sociable person, rather than sit at home alone, chose to enjoy the companionship of other young friends in the community. Even though the young husband had no information indicating any unfaithfulness, his imagination prompted him to worry lest his wife lose her love for him.

In my reply I advised this young husband to channel his imagination in a different direction. I urged that instead of allowing the screen of his brain to present pictures of the social tragedy that might come into his life, he use his imagination to devise means by which he could become a better husband and a better companion to his wife.

A young woman wrote, "I have just read an article which says we should not worry. But how can a person keep from worrying when she has personal difficulties and troubles which are discouraging? I fear that I cannot find a solution for my present problems."

I reminded this person that the remedy for worry does not consist so much in applying the negative approach of self-discipline (making a firm determination not to worry) as in using the positive approach by which one keeps his mind occupied with useful, profitable, and pleasant activities that crowd out the opportunity to worry.

The Christian philosophy provides the perfect antidote for worry, which the Master's words perfectly summarize: "Take no thought, saying, What shall we eat? or, What shall we drink? or, Wherewithal shall we be clothed? . . . For your heavenly Father knoweth that ye have need of all these things. But seek ye first the kingdom of God, and his righteousness; and all these things shall be added unto you. Take therefore no thought for the morrow: for the morrow shall take thought for the things of itself. Sufficient unto the day is the evil thereof." Matthew 6:31-34.

YOUR BRAIN IS BOSS

The human brain serves as headquarters for the most marvelous of all communications systems. Though weighing a mere three pounds, it receives a steady flow of coded messages from all parts of the body, which constitute the data out of which it constructs a perpetual inventory of what takes place both within and without the body.

Depending on what the sense organs register, the information coming from the outside may pertain only to the immediate surroundings or be so far-reaching as to report what takes place on the other side of the world or even in outer space.

Some of the circumstances reported to the brain require immediate handling. This the brain does by issuing directions to certain of the body's muscles or by making decisions which influence the person's conduct at some later time.

Other kinds of information brought to the brain provide information which the brain stores away in orderly fashion in the form of memories. These remain available for later use as occasion may require.

Other items reaching the brain merely round out its awareness of present conditions and do not require immediate action or the making of decisions. They are not important enough for storage in the form of memories. This third kind of information may be important, momentarily, as it indicates to the brain what is presently feasible with respect to the other matters taking place. A great deal of this background information is not even registered in conscious thought, but it has its influence, nonetheless, on other occurrences in the brain.

The CTC (Central Traffic Control) systems now in operation on many railroads illustrate how the brain serves as communications headquarters. It happens that the railroad traffic passing the town in which I live is regulated by such a system. This particular unit controls the movement of trains between Colton and Indio, California, a distance of eighty miles, and the freight traffic is particularly heavy over this stretch of the Southern Pacific line. Near the center a pass creeps between two of southern California's larger mountains, where the railroad attains an elevation of about 2,300 feet. Relatively steep grades for freight trains lead to this pass on both the east and the west.

The CTC headquarters for this eighty-mile stretch of track is located at Beaumont, at the summit. Here a control room with elaborate electronic devices receives and sends many signals per hour. Here a large map pictures each railroad siding, each switch, each company telephone, and each semaphore.

On the map many bulbs of different colors light up to indicate the position of each train in the eighty-mile stretch, the position of every switch whether open or closed, and the color of every signal light.

Close by the map an operator sits at a panel of electric connections which control all the switches and signal lights between Colton and Indio.

The operator keeps continuously informed of the position and movements of every train within the section of railroad for which he is responsible. Thus he directs the flow of traffic by the mere touch of the control buttons at his fingertips, so that one train shunts onto a siding and halts by a red light while another passes on the main line. Should an unexpected delay involve a westbound train, the eastbound train can be required to wait, by the mere closing of an electrical circuit, until the other safely moves off the main line. After a train has left the siding on which it has waited for passing trains, the operator in the control room closes the switch so that the brakeman riding on the train does not even have to touch his feet to the ground.

All the rail traffic along this eighty-mile stretch moves and stops only as directed from headquarters. When an engineer sees a green light, he knows he can safely make normal progress. With a yellow light staring him in the face, he must reduce the speed of his train in preparation for a stop or for moving onto a siding. When a red light appears, he dare not pass until the signaled instructions change.

The human brain resembles the control room just described, only many times more complicated. Always informed on present conditions both inside and outside the body, it makes decisions on the strength of this information, and sends instructions to one part or another of the body. But one feature in our illustration breaks down. In the control room at the CTC headquarters an *operator* interprets the signals that come under his observation and makes decisions by which he turns on a red light or a yellow light or opens the switch to a sidetrack. But in the human brain, there is no *operator*. And herein lies the difference between the brain and all of the marvelous electronic devices on which so many activities of our civilized way of life depend.

In order to further our understanding of how the brain

serves as boss both of the body and of the personality, we will notice that it consists essentially of three major parts: (1) the brainstem and the hypothalamus, which control many of the body's automatic functions; (2) the cerebellum, which deals with the coordination of muscle action; and (3) the cerebral cortex, in which the intellectual functions occur. These major portions of the brain closely interrelate so that they carry on their respective duties harmoniously. Just as in a business partnership there is a senior partner and a junior partner, with the senior partner having the controlling influence, so in the brain the cerebral cortex dominates the triumvirate.

For example, the brainstem—that portion of the brain situated deep in the back part of the braincase and which connects, through a large opening in the base of the skull, with the spinal cord—normally controls breathing. The so-called respiratory center of the brainstem monitors the body's need for oxygen, moment by moment, and balances against the action of the muscles which expand the chest and thus draw air into the lungs. When the need for oxygen increases, as when a person begins vigorous exercise, the control center for breathing not only causes the diaphragm and the muscles between the ribs to contract more forcibly, thus enabling the lungs to accommodate the larger volume of air, but it steps up the rhythm of breathing so that one breathing cycle follows another in faster succession. Similarly, when the body rests, the need for oxygen declines, and the respiratory center exerts the opposite influence so that the breathing becomes more shallow and less frequent.

As we have just said, the cerebral cortex can even give orders to the other parts of the brain. Continuing our reference to breathing, a person can decide that he wants to breathe more deeply or more rapidly, or both. Such a choice involves the cerebral cortex not only in making the decision, but also in overruling the respiratory center of the brainstem. Thus a person can control each breath he takes. He can even hold his breath until the tissues of his body become so much in need of oxygen that he loses consciousness. At such time the cerebral cortex, due to the shortage of oxygen, fails to function. At this point the respiratory center of the brainstem takes over again, and breathing resumes in time to save the individual's life.

We might say, then, that the usual control of breathing is automatic because the cerebral cortex entrusts this function to the brainstem. The cerebral cortex can, however, intrude and temporarily usurp the control of breathing; but when a person tires of giving conscious thought to each breath he takes, the control returns to the brainstem.

We have used the example of breathing to illustrate the manner in which the brainstem and hypothalamus care for the body's important functions. The rate of the heartbeat, body temperature, and the concentration of blood sugar are similarly controlled. Fortunately the brain is so organized that it automatically cares for many vital functions without the need for conscious supervision. This leaves the cerebral cortex free to concern itself, as far as a person's conscious thoughts are concerned, with matters other than the recurring needs of the organs and tissues.

Next let us notice how the cerebellum—the second of the three parts of the brain we have listed—carries on its work automatically but still under the supervision of the cerebral cortex.

The cerebellum deals with the activities of the body's various skeletal muscles—those muscles that move the arms, legs, fingers, and toes; the muscles of the face that provide facial expression; and the muscles of the neck, back, and abdomen—in fact, all of the muscles of the body except those that provide for movement within the organs and blood vessels. Every skeletal muscle of the body receives many nerve fibers which cause the muscle to contract, when necessary, and which regulate the strength of the muscle's contraction.

The cerebral cortex makes the decisions regarding the use of these muscles. Here a person determines when he wishes to walk and where he wants to go, but the cerebellum controls the details of the muscle actions involved in walking. Consider for a moment the apparently simple action of putting the right foot forward as in taking a step when walking. In order for the right leg to move ahead, those muscles that would cause the leg to move backward must relax. Thus certain muscles contract while others relax. The cerebellum cares for these details.

Perhaps we can get an even better idea of how the brain "bosses" the body's activities when we consider the rapid and intricate activities of a pianist's hands and fingers while playing a difficult passage of music. We recognize that practice develops skill, and sometimes we assume, incorrectly, that a pianist's practice periods train the *muscles* in his arms, hands, and fingers. Whereas the muscles do become stronger as the result of practice, the real advantage of practice is in developing the nerve circuits by which the brain controls the muscles of the arms, hands, and fingers.

In developing musical skill, the cerebral cortex does not have time to give individual direction to the action of each muscle as it contracts and relaxes. The musician's conscious thoughts in his cerebral cortex are concerned with the theme of the musical selection, with the expression which he uses in interpreting this theme, with the tempo of the music, and with the sequence of one musical phrase after another. The subordinate parts of the brain care for the selection of the proper keys on the piano.

Early in a musician's experience with a certain piece of music, he consciously considers the notes and chords involved in rendering the selection. As his skill develops, however, his cerebral cortex delegates these details of finger action and muscle coordination to portions of the brain which function unconsciously.

We have already characterized the human brain as the headquarters of the most elaborate communications network known to mankind. Modern knowledge of electronics gives us some insight into the means by which combinations of nerve circuits can give rise to the type of automatic control which the brain exerts over the other parts of the body. We oversimplify, however, to compare the human brain to an electronic computer. For example, the human brain contains around twelve or thirteen billion structural units (nerve cells)—five times as many nerve cells as people in the world. When we understand that the various possibilities of conscious thought and automatic controls are brought about by variable circuits made possible by changing relationships among the nerve cells, we see that the opportunities for creative thought, in addition to routine control of the body's functions and activities, are beyond computation.

We have noticed that the cerebral cortex makes decisions and creates intellectual activities. Higher animals have brains which in their anatomical features resemble the human brain. The brains of horses, dogs, monkeys, and many other animals all have a cerebral cortex. But the brains of the higher animals, excluding man, direct automatic activities; they do not deal with major decision making, with idealism, with creative undertakings, or with matters of intellectual advancement or moral significance.

Your Brain Is Boss

Only man, among the creatures formed at creation, has the power of choice, which he can use as a means of planning his personal future and of carrying out his plans at appropriate times. This power of choice, depending as it does on the normal functioning of the cerebral cortex, makes man a free moral agent. Because of this power of choice, the Creator holds man responsible for the use he makes of his life's talents and energies.

Because of man's ability to determine what he does and when, God has seen fit to make available to him the Ten Commandment law. When man chooses to live in harmony with this law, he reaps the benefits of God's approval, of personal blessings during the present life, and of eternal life in the future as offered to those who choose to take advantage of Christ's atonement for the human race. Using this same God-given power of choice, a man may choose to follow a course of action out of harmony with the provisions of God's law. In such a case, the unfavorable consequences he reaps still result from his exercise of free will.

Truly, then, the brain is boss, not only of a person's conduct, moment by moment and day by day, but also of his eternal destiny.

HEALTHY BODY, ACTIVE MIND

Medical literature says a great deal about the mind-body relationship. Indeed, an entire branch of medical science known as psychosomatic medicine has developed from the recognition that attitudes of the mind and emotional states influence the physical health. The previous chapter, "Your Brain Is Boss," explains that the nerve circuits enable the brain, as the dominant organ, to control the activities of all parts of the body.

In this chapter we shall reverse the emphasis on the relationship between the brain and the rest of the body and speak of the body's influence on the brain.

We all agree, I am sure, that the brain is the most important organ of the human body. The experience of consciousness by which the individual knows what occurs around him originates in the brain, which interprets sensations, records the memories of past events, and keeps them available for recall. The brain makes all decisions and controls all conduct in harmony with these decisions. The brain develops imagination and formulates plans for the future. It is in the brain that the higher intellectual processes take place—those that require the exercise of judgment and conformity to one's ideals and purposes. The brain establishes a personal philosophy and recognizes one's role in God's great plan.

Healthy Body, Active Mind

Inasmuch as a person's entire conscious experience depends upon what takes place in his brain, we can logically contend that the other organs of the body carry out their most important functions as they contribute to the welfare of the brain. For example, we say that the heart propels the blood throughout the body, but inasmuch as the brain consists, essentially, of billions of nerve cells which must remain alive and in good condition in order for the brain to operate normally, it becomes clear that the heart's most important function is to propel blood through the brain, thus satisfying its physical needs.

In similar manner, the most important function of the lungs is to take oxygen from the air so that a portion of it can go to the brain, where it contributes to the respiratory needs of the nerve cells.

The organs of digestion also subserve the brain by providing food materials which, when carried to the brain by the bloodstream, nourish its cells.

The organs of elimination also serve the brain as they remove from the circulating blood the waste products produced by cell activity throughout the body. Allowed to accumulate in the blood, these waste products would promptly curtail the functions of the brain.

Even the bones play their part as they encase the brain to protect it from ordinary injury. Other bones contribute by keeping the body intact and supporting it during locomotion.

The muscles make it possible for the body to move and thus carry the brain where it needs to go.

When the various organs of the body function at their best, the brain benefits by this healthy state and performs at peak capacity also. In this sense physical health promotes intellectual activity. Not that the person with the best physique is thereby the best scholar, but when a person in good physical health desires to use his brain efficiently and to develop his mental faculties to their ultimate, he holds a definite advantage over one who lacks physical vigor.

Also the reverse holds true, for factors which reduce a person's physical well-being limit his intellectual accomplishments. Notice the following list of conditions which lower one's physical vitality and at the same time limit his ability to think clearly and effectively.

1. Fatigue. Weariness from physical overexertion hampers creative thinking and concentrated study. The same factors that make it difficult to use the muscles handicap the brain also. When the body's energy ebbs, all of its organs slow down till such time as the energy is replenished.

2. Illness. Any form of illness, such as infection, handicaps the entire body. An emergency summons the natural resources of energy to resist infection or whatever other factor has caused the illness. Furthermore, the processes of healing require their share of vital energy. If in the course of the illness, toxins circulate in the blood or if the metabolism of the body's cells varies, the brain suffers reduced efficiency along with the other organs.

3. Reduced Blood Supply. Any circumstance that interferes with the normal supply of blood to the brain retards its functions. The simple occurrence of losing consciousness when a person faints illustrates the effect on the brain of a reduced blood supply. In cases of hardening of the arteries, the victim's mental activities become sluggish because the diseased arteries cannot carry as much blood as the brain requires to maintain a normal condition.

4. Dissipation. Indiscretions in a person's way of life cost a high price in terms of the expenditure of energy. Loss of sleep, overindulgence of appetite, sexual excesses, the use of intoxicating drinks, reduce one's general vitality and

thus proportionately handicap the brain.

5. Poisons and Drugs. These substances dissolve in the body's fluids and flood the tissues. Some poisons and drugs affect one organ more than another, and some even specifically affect the brain cells. Even though a drug may principally affect some other part of the body, it usually alters the body's general functions sufficiently that the brain suffers in one way or another. Some chemical substances supposedly increase the activity of the brain, making the individual more alert mentally, but the aftermath of such an influence moves in the opposite direction, so when the unfavorable effects of the chemical substance are included along with its supposed stimulating effect on the brain, the total picture grows very unfavorable.

6. Physical Discomfort. The body and the mind are so intimately related that the brain cannot function efficiently when some of the body's energies combat such factors as physical discomfort. Each individual has only a certain quantum of vital energy upon which to draw for both his physical and mental activities. When part of this energy is used to endure uncomfortable conditions, the amount of energy left over for mental activity is less than normal. Here we find the reason that physical torture sometimes accompanies a "brainwashing" experience. Under circumstances of extreme discomfort, a person cannot think clearly, nor is his judgment reliable.

Now that we have considered several ways in which unfavorable physical conditions can handicap the brain, let us turn our attention to methods for improving physical fitness and thus setting the stage for the most effective use of the brain. A brain operating in a healthy body is best capable of thinking clearly, of exercising good judgment, and of using imagination wisely in creative enterprises. You will notice that I list simple factors below—so simple, in fact, that you may succumb to the temptation to pass them by, thinking that such things really do not matter—for maintaining good physical fitness is not a complicated matter. It depends merely on following a plan similar to what a person uses when he wants to increase his assets at the bank. To build up one's credit there, he makes sure that the amount of money deposited day by day exceeds that which he withdraws.

Similarly, in maintaining good physical fitness, one must follow a way of life which conserves his vital energies rather than dissipates them. Every day a person derives a certain amount of energy from the food he eats. This compares favorably to making a deposit in the bank. His various activities tend to subtract from this total supply, and if he draws more liberally from his available supply of vital energy than his present balance of supply and demand justifies, his physical fitness reduces accordingly, thus making him more susceptible to illness and at the same time reducing the capacity of his brain for effective thinking.

In one area our comparison of maintaining a favorable bank balance does not truly represent the circumstances necessary to maintaining physical fitness, for if the comparison held true in all areas, then a person would build up higher and higher. In the matter of physical exercise, however, the reasonable use of the muscles is an asset rather than a liability. Physical exercise consumes a generous portion of the body's supply of fuel, but in the meantime appetite improves and additional food is desired. In the ultimate, physical activity carried on in a reasonable manner, day by day, increases one's store of vital energy. Continuing our analogy of the bank balance, we can liken it to an investment program by which the returns on the investment in-

Healthy Body, Active Mind

crease a person's assets over a prolonged period.

Now for our list of the ways in which a person can improve and maintain his physical fitness and thus keep his mental ability at its best:

1. *Choose a simple, adequate diet.* A person's "taste" for certain foods does not reliably call for the best foods. A person develops food preferences just like habits, and one's preferences depend more on his pampered desire for gustatory pleasure than on the body's basic needs.

This does not mean that all tasty foods are bad or that only the foods with objectionable flavors are nutritious. A person with a sensible appetite can derive just as much enjoyment from the simpler, wholesome foods as from indulging a perverted desire for highly seasoned and "exotic" foods.

One need not be a nutritionist to choose his food wisely. It does help, however, to look over some normal diet lists prepared by a person who has made a study of the body's nutritional needs. Other than this, three simple rules will help a person select his food wisely: (a) eat a relatively small variety of foods at any one meal; (b) vary the kinds of food eaten from day to day; and (c) give preference to fruits, vegetables, and whole-grain foods over the highly refined cereal products, rich foods with a high content of fat, and sweetened desserts.

2. *Eat regularly.* Between-meal snacks upset the leisurely rhythm of the digestive organs and thus set the stage for indigestion. In most cases going without breakfast or taking only a doughnut and a hot drink instead of a good breakfast makes it easy for a person to fall into the habit of irregular eating during the rest of the day.

3. *Abstain from personal indulgences.* Tea and coffee contain caffeine, a stimulating drug. Liquor contains alco-

hol, which temporarily depresses the brain's activities, reducing coordination and impairing judgment. Alcohol, in the long run, ruins the personality by tempting its user to evade life's stern realities. Tobacco contains nicotine, a poisonous drug, in addition to the "tars" resulting from its combustion, which irritate certain tissues and make them more susceptible to cancer. Smoking accelerates the aging of the tissues so that the smoker becomes vulnerable to certain types of heart disease, to degenerative diseases of the lungs, and to certain types of cancer at earlier ages than these ordinarily develop. Unwarranted use of sleeping pills and other drugs damages tissue and subtly undermines the user's general health and physical fitness.

4. *Sleep sufficiently.* Most adults require eight hours of sleep each night. Children and those in frail health need more. But the demands of strenuous living make it easy for a person to encroach on the time he should invest in the restoration of his body's resources. Staying up until eleven or twelve and then rising at six in order to get to work on time leave only six or seven hours to build up the nervous vitality which the previous day's activities have depleted. If one habitually reduces his amount of sleep, how else can the body redeem the time but by taking "time out" for illness or by subtracting it from one's life expectancy? All the while, of course, the individual's mental efficiency operates at a lower level than would be the case if his way of life contributed to his total fitness.

5. *Exercise consistently—preferably outdoors.* God designed both components of the human organism—the physical and the mental—for activity. Neither can reach its zenith of efficiency unless the other keeps active.

I do not mean that you must put your muscles to the stretch at the same time your brain solves a problem. One's general health and efficiency will improve, however, if within the span of each day he strenuously involves both the brain and the muscles in some form of activity. Constructive use of the brain vitalizes the various organs. Active use of the muscles for at least an hour out of the twenty-four improves the efficiency of the entire blood-circulating system, aids digestion, and favors the more complete elimination of tissue waste products.

Physical exercise, preferably in an outdoor setting, may include the element of recreation. Increasing the circulation of blood through the brain as well as providing a favorable setting while the thoughts wander from one pleasant topic to another reduces nervous fatigue and prolongs the ability to engage in concentrated thought.

6. *Maintain an attitude of trust and optimism.* It is no accident that most healthy people are cheerful and most sick persons grumble. It may be somewhat debatable, however, which causes which. Is a person cheerful because he is healthy or healthy because he is cheerful? Perhaps the relationship works both ways!

Here we are concerned with what you can do to promote your physical health and thus improve your mental efficiency. In addition to the six items listed, deliberately assume an attitude of trust and optimism. Even though this may sometimes require that you "put on the act" when you don't feel like it, the final effect will promote your own physical and mental welfare.

The body's organs operate more efficiently in a casual atmosphere of cheerfulness, and the autonomic nervous system mirrors the emotional state and influences the body's various organs accordingly.

HOW FAITH CAN KEEP YOU WELL

20 "This situation is hard on my ulcer," a young friend remarked when I urged him to break away from his present employment so that he could continue his education.

Apparently he noticed the quick glance I gave at his mention of "ulcer," for he added, "Of course I don't have an ulcer—yet. But in this condition of uncertainty, it will not take long for one to develop."

God organized the nervous system to keep the various organs of the body functioning harmoniously. The human body is a "tuned" mechanism in which no part operates independently of the others. The brain exerts the controlling influence and modifies the functions of the various organs to fit the needs of the moment.

However, the brain serves another function—that of permitting abstract thinking and creative imagery. This is the most important evidence of the superiority of the human being over other forms of life. Understandably, then, with the brain having two functions to perform, a person's thoughts and emotions often affect the organs.

Strong emotions, such as fear and hate, cause the organs to drastically prepare for a momentary emergency. When these strong emotions persist, the organs eventually weaken under the strain of the continuing state of preparedness for action, and various forms of functional illness and even organic disease may develop as a result.

The sober thoughts a person thinks, though not accompanied by violent emotions, can influence the body's state of affairs. Thoughts in conflict, a difficult decision and mental turmoil over an unanswered philosophical question, do their mischief by upsetting the delicate balances of nervous control which regulate the activities of the organs.

The question which probably most upsets a person and can do the most to undermine his state of health is the simple question Why?

A young child uses "Why?" as a tool to discover the world about him and to satisfy his curiosity regarding people, things, and relationships. And the quest continues through life. It impels a scientist to engage in research and a philosopher to probe the unknown. The student of history, of politics, or of human relations constantly seeks answers to the same question. A businessman bases his plans for investment or for expansion on the best answers he can find to this question.

But the question becomes most troublesome when a person examines his relationship to the sequence of life—present, past, and future. This question, applied to oneself, can either provide the motivation for major accomplishments or ruin one's peace of mind and, therefore, one's health.

In pursuing the question Why? the human mind meets its greatest challenge. And in finding the answers to this question as they relate to the individual, a person needs a prevailing, stabilizing influence against which he can measure himself and by which he can comfortably reconcile himself to his station in life and to the circumstances which he must face. Here faith comes into its own.

Let us now consider seven of the Why? questions with which a person has to wrestle. After listing the questions, we will go over them again to show how faith provides the only satisfactory answers.

1. Why am I who I am? This question relates to factors over which an individual has no control. As a child becomes old enough to evaluate his station in life, he may wish that he were of the opposite sex, that he had been born into a family which lives in a more favored part of town, or that he had the personal aptitudes which would make him outstanding in a field of his own choosing.

2. Why was I born at this particular time? Most people admire the technical progress of our modern times, so they rejoice to live in this era rather than at some previous time. Some, however, feel confused and overwhelmed by modern complexities, and moan, "I wish I could have lived in 'the good old days.'"

3. Why am I here? This question takes on broad meaning for those who indulge in philosophical speculation on the purpose of life. Those with abundant ambition come under the spell of a desire to accomplish great things and make their lives really worthwhile.

4. Why do unwelcome circumstances come to me? Many take pride in closely planning their lives. Plans are excellent and help a person use his advantages and energies effectively, but sometimes plans miscarry. Illness, misfortune, hardship, lack of appreciation, displacement by a rival, unexpected obligations, unhappiness at home, or failure in some cherished endeavor—any of these may prevent a person from fulfilling his plans or realizing the benefits he had expected to enjoy. Then the question Why? becomes colored with resentment for life's inequities.

5. Why does God forbid some of the things I like to do? Wrestling with one's conscience ruins peace of mind, and can, over long periods, adversely affect general health.

How Faith Can Keep You Well

6. Why does God permit wickedness? Law-abiding people ask this question. They know that God's law forbids base conduct, and they question His equity in dealing with mankind when they observe that the dishonest, cruel, and immoral seem to prosper more in life than others with exemplary conduct.

7. Why does God not answer my prayer the way I want Him to? Persons who believe in the efficacy of prayer but who have tried to use prayer as a convenient means of realizing selfish desires pose this question.

The questions just listed cause people to become confused, fill their minds with doubt, and thus promote poor health. Fortunately, God does not leave us without satisfactory answers. It requires faith, however, to accept and apply in daily living the answers He gives.

Once a person develops the degree of faith which enables him to accept the answers, the conflict in his thinking ceases. He becomes reconciled to God's pattern of dealing with the human race, and this reconciliation eliminates mental turmoil. Faith enables a person to trust God to direct the affairs of his life. Confidence in God's dealings removes the anxiety, the fear, and the doubts which rob a person of vitality and health.

Let us now repeat the questions and notice how the Bible provides the answers.

1. Why am I who I am? In Psalm 139 we find David's statement that God knew the individual even before birth and that He supervised the marvelous events of human development: "For thou hast possessed my reins: thou hast covered me in my mother's womb. . . . My substance was not hid from thee, when I was made in secret, and curiously wrought in the lowest parts of the earth [the womb]. Thine eyes did see my substance, yet being unperfect; and in thy

book all my members were written, which in continuance were fashioned, when as yet there was none of them." Verses 13-16.

2. Why was I born at this particular time? In the fourth chapter of the Book of Esther, verse 14, Mordecai counsels his cousin, Queen Esther, "Who knoweth whether thou art come to the kingdom for such a time as this?"

Mordecai clearly perceived that this young woman was born at the right time to fulfill God's purpose in molding the affairs of the kingdom of Persia for the protection of His chosen people. The same God who prearranged the life and circumstances of Queen Esther will do this for every individual who, by faith, submits to God's way in his life.

3. Why am I here? Paul in 1 Corinthians 10:31 provides our answer to this question: "Whether therefore ye eat, or drink, or whatsoever ye do, do all to the glory of God."

We exist for only one purpose—to glorify God. Furthermore, we honor His name as we order our lives in harmony with His purpose for us. The divine plan is not a mandate, however, for it remains with the individual to accept or reject God's purposes for him.

4. Why do unwelcome circumstances come to me? Our earth—the utopia of creative Infinite Intelligence—has degenerated under destructive, diabolical cunning. Paul in his letter to the Romans describes the whole creation as enslaved to purposelessness and groaning in pain. (Romans 8:20, 22.) All disaster thus originates with an antagonistic power —the antithesis of God—the devil. Yet God thwarts Satan's onslaughts against the person who has committed his life in trust to Him. "And we know that all things work together for good to them that love God." Romans 8:28. Thus God brings good out of the unwelcome tragedies of life, and through them we, by maintaining a loving faith in God,

learn obedience and develop spiritual maturity. (Hebrews 12:11.)

5. Why does God forbid some of the things I like to do? This question brings up a consideration of God's law—the rule of conduct by which He indicates the type of behavior which He accepts.

Early in the Book of Romans the Apostle Paul mentions the importance of faith when he says, "The just shall live by faith." But he devotes most of the Book of Romans to a discussion of God's law. In chapter 3, using an interesting figure of speech, Paul tells the function of the law: "It is the straightedge of the Law that shows us how crooked we are." Romans 3:20, Phillips.*

God is our Creator and Judge, and we must not challenge His wisdom in stating, through His law, the type of conduct He requires.

6. Why does God permit wickedness? This question bothered King David, for we notice in Psalm 73:3 that he wrote, "I was envious at the foolish, when I saw the prosperity of the wicked." David observed that many times the wicked escape punishment and seem favored above those who abide by God's law.

God permits us to exercise our power of choice. Each individual can accept or reject God's plan for his life. In Hebrews 11 we have an insight into what this meant in the experience of Moses.

Moses, a child of Providence, had a great task God planned for him to do. However, Moses could have turned his back on God's plan and chosen to follow his personal inclinations. The text in Hebrews describes his choice: "By faith Moses, when he was come to years, refused to be called

---
*From The New Testament in Modern English. Copyright, J. B. Phillips, 1958. Used by permission of The Macmillan Company.

the son of Pharaoh's daughter; choosing rather to suffer affliction with the people of God, than to enjoy the pleasures of sin for a season." Hebrews 11:24, 25.

Many do otherwise than Moses did, preferring "to enjoy the pleasures of sin." God in His mercy permits this, reserving for the day of final judgment the penalty or the reward.

7. Why does God not answer my prayer the way I want Him to? The more we learn about God and His character, the more we realize that He is merciful and that His interests bring to His human children the greatest opportunities for development in harmony with the divine purpose. God knows our motives when we pray and translates our prayers in ways that will bring to us the greatest eternal benefits rather than catering to our whims and selfish desires of the moment. The prophet Isaiah summarized this principle when he said, "Your iniquities have separated between you and your God, and your sins have hid his face from you, that he will not hear." Isaiah 59:2. In other words, our selfishness and perversity prevent God from answering our prayers in the ways we would prefer. When we pray as did the Master, "Not as I will, but as thou wilt" (Matthew 26:39), we can rest assured that God will hear and answer in the best way.

Now that we have observed the Bible's answers to the questions that disquiet the human mind and make the body susceptible to disease, let me emphasize that the exercise of faith enables a person to apply these answers in his own experience. By such an exercise one's mind becomes at peace and his wholesome mental attitude sustains his health.

"Don't worry over anything whatever; tell God every detail of your needs in earnest and thankful prayer, and the peace of God, which transcends human understanding, will keep constant guard over your hearts and minds as they rest in Christ Jesus." Philippians 4:6, 7, Phillips.

THE SIGNIFICANCE OF BEING A PERSON

 Name any city that you choose—San Francisco, New Orleans, Chicago, Detroit, Boston, or Denver—then take a close look, and you find that each is unique, with its own unusual features.

Consider Boston as an example. It holds a definite charm for those who admire its particular characteristics. Boston's individuality depends on much more than its geography with the Charles River flowing past Beacon Hill to enter the Boston Harbor. Its uniqueness embraces more than the Boston Common, Faneuil Hall, the Old North Church, the unusual situation of its airport, and its striking new Prudential Tower.

It takes all of these tangible characteristics plus the human qualities of its people—their way of thinking, their likes and dislikes, and their customs—to give the city its "personality."

Human beings are the same way. Every individual has qualities that set him apart from every other person. True, each human being has a heart, a pair of lungs, a liver, a stomach, and a brain; every person has a nose, a mouth, two ears, and two eyes. Yet, though everyone possesses similar visible features, every person also differs.

One may stand the same height as another, have the same complexion, and may even have the same type of square shoulders, yet not only do his fingerprints differ from others, but also his gait, his way of wrinkling his forehead, and his manner of holding a fork.

The blending of those intangibles which account for the way one thinks and feels, the way he responds with love or hate, and the way he relates to those whose lives flow next to his contributes to a person's individuality even more than do his unique physical characteristics.

A city has sidewalks and streets so that people can go and come. Gas mains, water mains, and power lines bring the utilities to every building in town. Sewers and disposal systems take away the refuse. Hotels provide facilities for lodging and recreation. Cafés and dining halls feed the people. Schools, museums, and libraries foster education and culture. Communication networks for telephone, radio, and TV enable those in the city to keep in touch with one another and even to give and take information to and from other cities and distant parts of the world.

Yet not one of a city's facilities is adequate of itself. The city functions smoothly only as each facility works harmoniously with the others and provides its share of service. In order to avoid confusion, the various services must coordinate.

Every city provides for such coordination by maintaining a city government. A city hall houses the mayor's office and the quarters for other city officials. There is a chamber where the city council meets to discuss the city's affairs and to make decisions for its operation. This central control by the mayor and the city council bears the same relationship to the activities of a city as the brain bears to the functions of the human body.

No doubt as you read the earlier chapters of this book, you kept anticipating those later chapters which describe the brain, its activities, and its relationships to the other

The Significance of Being a Person

organs. Probably you hoped that when you came to these later chapters you would find answers to your questions, What is consciousness? and What is personality? You were probably disappointed, then, when on reaching these chapters you discovered that we can point to no single site in the brain as the seat of consciousness nor to a section responsible for personality.

Consciousness is simply an awareness by which a person knows what happens in his surroundings. It depends upon the normal operation of all parts of the brain and of the nerves that enter and leave the brain. Similarly, the important human qualities of personality and character are not limited to any one part of the brain but are determined and governed by all parts of the brain and body, functioning unitedly and cooperatively.

Many of the activities of the brain defy analysis. Scientists have long tried to fathom the mysteries that underlie the human ability to remember, to experience imagination, and to make decisions. We often compare the brain to a computer, but we cannot understand its circuitry or say, This is the area in which a person's conscience resides.

We cannot identify the important features of the personality in the laboratory. An examination of the brain of a departed scientist does not indicate that he possessed any more intelligence than a person who had finished only eight grades of school. We know of no anatomical reason why one person is honest and reliable whereas another is shiftless and undependable. Even the electron microscope reveals nothing in the brain tissues which would explain the difference between two persons, one gentle and loving and the other rude and selfish.

The human body is not a mere mechanical device. Even though the Creator built into it many automatic controls and numerous margins of safety, still it cannot indefinitely perpetuate itself. Creative power designed the human body in the first place. A constant flow of this same supernatural power maintains the body's functions now.

The greatest evidence that the human body is more than a machine is the brain's ability to exercise the power of choice. A machine can only follow commands. Even with automatic features, it has no capacity to decide. When one control button is pushed, it performs whatever activity this particular electric circuit dictates. Only as the operator pushes some other control button does a machine perform differently than before.

Not so with the human organism. Human motivations develop from within. The capacity to decide, the ability to choose and then to follow the mapped-out plan of conduct, is the greatest gift a beneficent Creator has bestowed on human beings. Only man possesses this ability to determine his own destiny.

The study of the design of the human organism should magnify your responsibilities as a person. By your own choice and by the exercise of your own determination, you may use your body and your mentality in profitable ways, or you may squander your opportunities and permit your life to fall short of its possibilities. You must determine which.

Apparently the thoughts of the lawyer, whose conversation with Christ Matthew recorded in his Gospel, had run along these lines. When he asked for guidance in life's most important matters, the Master pointed out the great value of love—love to God and love to "thy neighbour."

True love is not simple ecstasy. It is the most complex experience of which the brain is capable. It binds up all of one's resources of intellect as well as emotion.

The Significance of Being a Person

In becoming master of your own destiny, choose to cultivate love—the kind that manifests itself in loyalty to God and adherence to the principles of Christian living which the Scriptures enunciate. Then you will reflect the love of God in your dealings with your fellowmen, and life will be more rewarding, both to you and to those who come under your influence.